Security of the China Pakistan Economic Corridor (CPEC)

This book analyses the strategic and economic significance of the China-Pakistan Economic Corridor (CPEC), with a particular focus on the region of Balochistan. Given the history of multiple insurgencies that the Pakistani Military has confronted in Balochistan, the book examines the region's intricate conflict ecosystem and security landscape, which poses potential threats to the CPEC.

Structured chronologically, the book traces the evolution of the Pakistani Army's counterinsurgency practices inherited in 1947 from the British Indian Army's culture of fighting *small wars* through to the contemporary counterinsurgency (COIN) adaptation in the "war on terror", and afterwards, to the fifth round of the Balochistan insurgency. The analysis centres on the development of counterinsurgency theory and practice by the Pakistani Army. It empirically investigates the efficacy of the COIN strategy in Balochistan. The author argues that the approach significantly changed after conceptualising the doctrine, especially from 2016 onwards, from "butcher and bolt" to the inclusion of critical components like political primacy, affect-based, and focused use of force, "winning hearts and minds" and rules of engagement. As a result, there was reduced violence and an increased number of insurgent surrenders. This book concludes that the Pakistani Army has largely controlled the insurgency in Balochistan. However, simultaneously, there is an urgent need to reduce tangible support to the insurgents through porous borders and implement an effective strategy to sever the nexus between the Islamic State of Khorasan (ISKP) and the Baloch insurgent organisations, as well as the sectarian militant organisations in Balochistan province. This is crucial to ending the insurgency and ensuring the security of CPEC.

A novel contribution to the study of counterinsurgency and the importance of CPEC to China's foreign policy and diplomacy, as well as its effects on the conflict dynamics in Balochistan, the book will be of interest to researchers studying War and Conflict Studies, Terrorism Studies, International Relations, Security and Strategic Studies, and South Asian and Chinese Studies.

Khurram Shahzad Siddiqui is Assistant Professor of War Studies in the School of Social Sciences and Humanities at the University of Management and Technology (UMT) Lahore, Pakistan.

Routledge Studies in South Asian Politics

For more information about this series, please visit: www.routledge.com/asianstudies/series/RSSAP

"This is a highly original study of the political and security impact of CPEC in Balochistan. The author has a forensic socioeconomic understanding of the region as well as Pakistan and China's wider security policies and interests. His appreciation of the drivers of the Baloch insurgency, and the Pakistani military's response to that conflict, is compelling and convincing. This book is a major contribution to South Asian studies; it throws much-needed light on the complex effects of one of China's most important and costly investments in the region. It will also be of significant value to scholars of terrorism and insurgencies."

Dr. Edward Burke, *Assistant Professor in War Studies,*
University College Dublin, Ireland

Security of the China Pakistan Economic Corridor (CPEC)

Counterinsurgency in Balochistan

Khurram Shahzad Siddiqui

Routledge
Taylor & Francis Group

LONDON AND NEW YORK

First published 2024
by Routledge
4 Park Square, Milton Park, Abingdon, Oxon OX14 4RN

and by Routledge
605 Third Avenue, New York, NY 10158

Routledge is an imprint of the Taylor & Francis Group, an informa business

British Library Cataloguing-in-Publication Data
A catalogue record for this book is available from the British Library

ISBN: 978-1-032-53839-6 (hbk)
ISBN: 978-1-032-53838-9 (pbk)
ISBN: 978-1-003-41390-5 (ebk)

DOI: 10.4324/9781003413905

Typeset in Times New Roman
by KnowledgeWorks Global Ltd.

To my parents who taught me how to persist in the face of seemingly insurmountable challenges.

And to

Nitasha, a beacon of affection's bliss,

Abdullah and *Rania*, my love that can't be missed.

Together, we create a masterpiece of love's embrace,

Eternal partners, woven in heart's sacred space.

Contents

Figures and Table

Figures

Table

Abbreviations/Acronyms

3PCM	Kilcullen's Three Pillars of Counterinsurgency Model
Af-Pak	Afghanistan-Pakistan
ANP	Awami National Party [a political party in Pakistan]
APP	Associated Press of Pakistan
ASWJ	Ahle-Sunnat-wal Jamaat [sectarian militant organisation in Pakistan]
B TEVTA	Balochistan Technical Education and Vocational Training Authority
BAP	Balochistan Awami Party [currently ruling Political Party in Balochistan]
BCIM	Bangladesh-China-India-Myanmar (Economic Corridor)
BE	Balochistan Education (Project)
BLA	Baloch Liberation Army [insurgent organisation in Balochistan]
BLUF	Baloch Liberation United Front [insurgent organisation in Balochistan]
BNP	Balochistan National Party [a political party in Balochistan]
BRA	Baloch Republican Army [insurgent organisation in Balochistan]
BRAS	Baloch Raaji-Aajohi e-Sangar [insurgent organisation in Balochistan]
BRG	Baloch Republican Guard
BRI	Belt and Road Initiative
BSSOs	Broad Spectrum Security Operations
C&SC	Command & Staff College
CAF	Civil Armed Forces
CCPO	Chief Capital Police Officer
CGS	Chief of General Staff
CIMIC	Civil-Military Coordination
CISAC	Center for International Security and Cooperation [Stanford University, US]
CJCSC	Chairman Joint Chiefs of Staff Committee
CM	Chief Minister

CMEC	China-Myanmar Economic Corridor
CMSR	Twenty-first Century Maritime Silk Route Economic Belt
COAS	Chief of the Army Staff
CPEC	China Pakistan Economic Corridor
D&E Directorate	Doctrine & Evaluation Directorate [Pakistani Army]
DCC	Defence Committee of the Cabinet
DDR	Disarmament, Demobilization, and Reintegration
DG	Director General
DW	Deutsche Welle
ECE	Early Childhood Education (Program)
ETIM	East Turkestan Islamic Movement
FATA	Federally Administered Tribal Areas
FC	Frontier Corps
FDI	Foreign Direct Investment
FF	Frontier Force
FIR	First Hand Information Reports
FM 3-24	Counterinsurgency Field Manual (US Army)
FWO	Frontier Works Organisation
GEOINT	Geospatial intelligence
GHQ	General Headquarters [Pakistani Army]
GOC	General Officer Commanding
GS Branch	General Staff Branch [Pakistani Army]
GSP	General Staff Publications [Pakistani Army]
GWOT	Global War on Terror
HUBCO	Hub Power Company Limited
HUMINT	Human Intelligence
IBO	Intelligence based Operations
ICC	International Cricket Council
ICJ	International Court of Justice
ICTO	Information, Communication & Technology Officer [Pakistani Army]
IDPs	Internally Displaced Persons
IED	Improvised Explosive Devices
IGT&E Branch	Inspector General Training and Evaluation Branch [Pakistani Army]
IHDs	In-House Discussions
IMF	International Monetary Fund
IO	Information Operation
ISI	Inter-Services Intelligence [Agency]
ISKP	Islamic State of Khorasan
ISPR	Inter Services Public Relations [Directorate]
ISR	Intelligence, Surveillance, and Reconnaissance
JuA	Jamaat-ul-Ahrar [sectarian militant organisation in Pakistan]
KPK	Khyber Pakhtunkhwa [One of the Provinces of Pakistan]
KSNP	Kalat State National Party

LEAs	Law Enforcement Agencies
LeB	Lashkar-e-Balochistan [insurgent organisation in Balochistan]
LeJ	Lashkar-e-Jhangvi [sectarian militant organisation in Pakistan]
LIC	Low Intensity Conflict
Lt Col	Lieutenant Colonel
MBS	Muhammad bin Salman [Saudi Crown Prince]
MDGs	Millennium Development Goals
MI	Military Intelligence
MO	Military Operations
MoU	Memorandum of Understanding
MT Directorate	Military Training Directorate
n.d.	no date
NAB	National Accountability Bureau [Pakistan]
NACTA	National Counter Terrorism Authority [Pakistan]
NAP	National Action Plan
NCHR	National Commission for Human Rights [Pakistan]
NDU	National Defence University
NFC	National Finance Commission
NGO	Non-governmental Organisation
NTF	Naval Task Force
NWFP	North West Frontier Province [now known as Khyber Pakhtunkhwa]
OBOR	One Belt, One Road
OCRs	Operation Completion Reports
OSA	Official Secret Act
PAD	Pakistan Army Doctrine
PFF	Punjab Frontier Force
PIF	Punjab Irregular Force
PIT	Pre induction Training (Cycle)
PM	Prime Minister
PMA	Pakistan Military Academy
PML-N	Pakistan Muslim League (Nawaz) [a political party in Pakistan]
PML-Q	Pakistan Muslim League (Quaid-e-Azam) [a political party in Pakistan]
PPP	Pakistan People's Party [a political party in Pakistan]
PRO	Public Relations Officer
Psy Ops	Psychological Operations
PTI	Pakistan Tehreek-e-Insaf [a political party in Pakistan]
R&D Directorate	Research & Development Directorate [Pakistani Army]
RAF	Royal Air Force
RCDS	Royal College of Defence Studies
RMA	Revolution in Military Affairs
ROE	Rules of Engagement
SATP	South Asia Terrorism Portal

SCO	Sub-Conventional Operations
SCW	Sub-Conventional Warfare
SI&T	School of Infantry & Tactics [Pakistani Army]
SIGINT	Signals intelligence
SITREPS	Situation Reports
SOPs	Standard Operating Procedure
SREB	Silk Road Economic Belt
SSD	Special Security Division [Pakistani Army for protecting CPEC]
SSG	Special Services Group [Pakistani Army]
SSP	Sipah-e-Sahaba [sectarian militant organisation in Pakistan]
TEWTs	Tactical Exercises Without Troops
THMCN	Trans-Himalayan Multi-Dimensional Connectivity Network (between Nepal and China)
TSDs	Tactics/Techniques, SOPs, and Drills
TTP	Tehreek-e-Taliban Pakistan
UAVs	Unmanned Aerial Vehicles
UBA	United Baloch Army
UCDP	Uppsala Conflict Data Program
UNDP	United Nations Development Programme
UNODC	United Nations Office on Drugs and Crime
VBMP	Voice of Baloch Missing Persons [Organisation in Balochistan]
VCOAS	Vice Chief of the Army Staff
WHAM	Winning Hearts and Minds (initiatives)

Acknowledgments

I am grateful to God for bestowing me with the intellect, vision, and strength to complete this ardours task of writing a book on a subject which is highly under researched. The world has no insight into the internal dynamics of the conflict eco-system of Balochistan, its implication on the CPEC, and how the Pakistani Army's institutional learning has shaped Balochistan's security environment.

I accumulated many debts of gratitude over the long journey of this book, which began as a PhD thesis researched between 2017 and 2019 and was completed in 2020, but subsequently revised and updated with further research. First of all, I owe a big thanks to Andrew Mumford and Katharine Adeney for supervising my research at University of Nottingham (UoN), UK. Andrew Mumford's invaluable academic support and encouragement always helped me immensely during this research. He also explained to me the value of publishing my research work as a book. I am also extremely grateful to Katharine Adeney for her insightful feedback, unwavering support and thorough scrutiny of my research work. Without the dedicated support of my supervisors, the research would not have been completed. I truly appreciate the significant contribution of the School of Politics and International Relations (SPIR), UoN for sponsoring my field research work in Pakistan, through the Postgraduate Research Fund, along with the Asia Research Institute.

I am incredibly thankful to Edward Burke and Bettina Renz, the established scholars and faculty members at the SPIR, UoN, who were very supportive and extremely helpful in my research. I am also thankful to Veronika Poniscjakova, Abdul-Jalilu Ateku, and Dishil Shrimankar for their assistance during my academic journey at Nottingham. I also thank my peers Nadia Al-Dayel, Luciano Pollichieni, and Khalid Jarral. I cannot forget Ilona Dane, my great friend from Derby, who always admired me, appreciated my research, and motivated me through difficult times.

During my fieldwork in Pakistan, I incurred many debts, and it would not be possible to name and thank everyone here. I am immensely grateful to all of my interviewees, including Rahimuddin Khan, who recently passed away. His insights and observations about military operations in East Pakistan and Balochistan, frequently referred to in the book, have been invaluable. I am very grateful to all the academics, police and military officers, civil servants, members of the civil society of Balochistan, and politicians who have been willing to share their views with me.

I owe a huge debt of gratitude to my comrade and friend, Dr. Muhammad Shoaib Pervez, a great source of inspiration and mentorship for me. He is an established scholar of International Relations. During my tenure as an Assistant Professor at UMT, he greatly motivated me to publish my research. He has always been candid in providing feedback on my book's manuscript. I also sincerely thank Dorothea Schaefter and her team at Routledge for their consistent support throughout the book's publication.

I am thankful to all my friends in Deutschland for their moral support. These include Dr. Somayeh Mohammadi, and Clara Evans.

I must highlight a few names of uniformed army officers to whom I owe my gratitude. This extremely rare species of officers helped me understand, both in theory and practice, the concepts of conflict, war, security, and defence. In their company, I truly grasped the profound meanings and significance of terms like *Comrades-in-arms* and *Band of Brothers*.

Foremost are Nasim Anwar (Frontier Force (FF) Regiment), Asim Suleiman (Corps of Engineers), and Abrar Ahmed (Punjab Regiment), who set the firm foundation of my understanding of military tactics and strategy.

I am thankful to Asif Ghafoor (Corps of Artillery), who emphasised the value of understanding the dynamics of sub-conventional conflict, including its theoretical underpinnings and practical manifestation, while advancing towards *Loe Sam* during *Operation Sherdil* (Bajaur Agency-FATA) in August 2008. His advice served as the initial impetus for me to conduct an in-depth study and critically analyse the sub-conventional warfare campaigns of the Pakistani Army. This book represents the culmination of my pursuit of knowledge and analysis of the sub-conventional operations conducted by the Pakistani Army in the more recent time period. I am equally grateful to Abid Mumtaz (Corps of Artillery), whose resolve as a commander during *Operation Sherdil* (Bajaur Agency – FATA) left deep imprints on me. Leading by his example, he demonstrated the qualities that answer the question, "*What are Generals Made of?*"

I am also grateful to Muhammad Feyyaz (Punjab Regiment), a soldier-scholar holding doctorate from Queen's University Belfast, for inspiring me through his published research on Terrorism Studies in prestigious academic journals. His mentorship has greatly aided me in pursuing my research endeavours. In the same category of soldier-scholar, I acknowledge Ahsan Illahi's (Punjab Regiment) enormous contribution in instilling in me a passion for military learning and the courage to write the truth despite all odds. I also owe thanks to Adeel Kazmi (Baloch Regiment) for always encouraging me in all my research endeavours. I must not forget to mention my very dear friend Hassan Abid (Corps of Artillery). He exhibited great valour and sacrificed his life during *Operation Al-Mizan*, at Jani Khel (Frontier Region Bannu) on February 8, 2011. He always encouraged me in all my academic and professional endeavours. May his soul rests in peace.

On a personal note, I am incredibly grateful to Tanzeela Anbreen, my elder sister, a great scholar of Applied Linguistics with a doctorate from the University of Bedfordshire, UK. She had been a great source of motivation for all my intellectual pursuits. I had frequent intellectually stimulating sojourns regarding my research at her place in central London. These discussions highly helped me in refining my research work. I am also thankful to her husband, Asim Qayyum, a second-generation military officer with high standards of traditional chivalry along with intellectualism. His insights into the various aspects of institutional learning of the military were a great help to my research.

I want to thank my parents, Shamim Ahmad Siddiqui and Tabasum Farhat, both retired civil gazetted officers with professional qualifications as an economist and an educationist, respectively. My parents inculcated in me a love for books and a passion for excellence. They taught me high morals and social values. They have always shown great interest in my book publication. I am fortunate to have my parents' unwavering moral and social support.

I am equally thankful to Tanzeel, Fatima, and Bazegha for their moral support to me during my academic journey of excellence. I ought to acknowledge the lovely gestures of hospitality from my nephew Moatisim (Momi) and my nieces Ifrah and Mustabshera (Mani) whenever I visited their home in central London during my research. Moreover, these kids have been very generous in not disturbing their mom (Dr. Tanzeela Anbreen) when she was engaged in long academic discussions with me, a very frequent occurrence during my research period in the United Kingdom.

I am incredibly grateful to my lovely wife, Nitasha, a medical doctor, for her love, patience, prayers, and unstinting and unconditional support to me. Along with her, I am thankful to my son Abdullah and daughter Rania who are the apples of my eyes and my raison d'être, Alhumdulilah. As I embark on this journey of writing and sharing my thoughts with the world, I can't help but envision a bright and promising future for my beloved children. May they continue to flourish, grow in wisdom and intellectualism, and find boundless success and happiness in all their endeavours, enriching the lives of others and making a positive difference in the world. Aamin.

Cast of Characters

Abdul Malik Baloch	Chief Minister of Balochistan (2013–2015)
Abdul Quddus Bizenjo	Chief Minister of Balochistan (2021-present)
Akhtar Mengal	Chief Minister of Balochistan (1997–1998), Political leader and chairman of Balochistan National Party (Mengal)
Allah Nazar Baloch	leader of the Balochistan Liberation Front (an armed group in Balochistan)
Ashfaq Parvez Kayani	Chief of Army Staff (2007–2013)
Asif Ali Zardari	President of Pakistan (2008–2013)
Asif Ghafoor	Director General of the Inter-Services Public Relations (ISPR) (2016–2020), Commander Southern Command & XII Corps (Quetta-Balochistan) (2022-present)
Asim Munir	Chief of Army Staff (2022-present)
Asim Saleem Bajwa	Director General of the Inter-Services Public Relations (ISPR) (2012–2016), Commander Southern Command & XII Corps (Quetta-Balochistan) (2017–2019), Chairman of the CPEC Authority (2019–2021)
Brahamdagh Khan Bugti	Baloch nationalist and leader of Baloch Republican Army (BRA) (an armed group in Balochistan) living in self-imposed exile in Switzerland
Hyrbyair Marri	Baloch nationalist and leader of Balochistan Liberation Army (BLA) (an armed group in Balochistan) living in self-imposed exile in London
Imran Khan	Chairman of Pakistan Tehreek-e-Insaf, Prime Minister of Pakistan (2018–2022)
Jam Kamal Khan	Chief Minister of Balochistan (2018–2021)
Kulbhushan Jadhav	Indian naval officer and spy arrested inside Balochistan in Mashkel near the border region of Chaman on 3 March 2016
Mian Raza Rabbani	Chairman of the Senate of Pakistan (2015–2018)
Mir Ahmad Yar Khan	Khan of Kalat (1933–1955)
Muhammad Shehbaz Sharif	Prime Minister of Pakistan (2022-present)

Muhammad Zia-ul-Haq	Chief of Army Staff and President of Pakistan (1978–1988)
Nawab Akbar Bugti	Chieftain of Bugti Tribe in Balochistan (1947–2006), Governor of Balochistan (1973, 1974), Chief Minister of Balochistan (1989–1990)
Nawaz Sharif	Prime Minister of Pakistan (1990–993; 1997–1999; 2013–2017)
Pervez Musharraf	Chief of Army Staff (1998–2007), Chief Executive of Pakistan (1999–2002), President of Pakistan (2001–2008)
Qamar Javed Bajwa	Chief of Army Staff (2016–2022)
Raheel Sharif	Chief of Army Staff (2013–2016)
Rahimuddin Khan	Governor of Balochistan (1978–1984), Chairman Joint Chiefs of Staff Committee (1984–1987)
Sarfraz Ali	Commander Southern Command & XII Corps (Quetta-Balochistan) (2020–2022)
Shujaat Zamir Dar	Inspector General, Frontier Corps Balochistan (2004–2007)
Wen Jiabao	Premier of the People's Republic of China (2003–2013)
Xi Jinping	President of the People's Republic of China (2013-present)
Yousaf Raza Gillani	Prime Minister of Pakistan (2008–2012)
Zulfikar Ali Bhutto	President of Pakistan (1971–1973), Prime Minister of Pakistan (1973–1977)

1 Introduction

Insurgencies are the by-products of interstate rivalries and the internal dynamics of a country (Gates & Roy, 2014, p. 191). This is very true in the case of Pakistan. The country has a rich history of dealing with insurgencies since its inception in 1947 after the partition of British India. The first-ever insurgency surfaced in the country back in April 1948 as a result of the forced accession of the state of Kalat (Balochistan) with Pakistan. Thereafter, the Pakistani Military fought insurgencies across the length and breadth of the country starting from erstwhile East Pakistan (now Bangladesh) to various rounds of insurgency in Balochistan, in the Federally Administered Tribal Areas (FATA) and the province of Khyber Pakhtunkhwa (KPK). There were mixed and diverse reasons for the surfacing and resurfacing of insurgencies in Pakistan. For instance, the East Pakistan insurgency in 1971 was the result of poor management, discriminatory and marginalising economic and social policies of the federation towards Bengalis along with the subjugation of the Bengalis by the use of force. Later, the interstate rivalry of Pakistan with India played a very significant role in the defeat of the Pakistani Army in erstwhile East Pakistan and led to the creation of independent Bangladesh. In the case of FATA, the factor which served as a stimulus to the rise of the insurgency was the official "anti-Taliban" stance, in the wake of US pressure after 9/11, adopted by the then Pakistani Government. This conflicted with the tribal ethos of allegiance to the Jihadis in FATA. Out of these varied phenomena of insurgencies, the Balochistan insurgency is unique as it has resurfaced five times in the past seventy years in the same geographical setting in the country. There are multiple reasons for the insurgency which can be broadly categorised as the deprivation of the Baloch people, the discriminatory economic policies of the federation towards Balochistan, the use of force by the military, the lack of control of Balochistan's natural resources (like natural gas, coal, etc.) by the province, and a continuous policy of subjugation of the Baloch people. Another glaring factor which has added an international dimension to the ongoing Balochistan insurgency in 2006 is the geo-strategic importance of the deep seaport of Gwadar and later in 2013, the Chinese Belt and Road Initiative (BRI) to get access to the Indian Ocean by linking the China Pakistan Economic Corridor (CPEC) to the port of Gwadar in Balochistan. The Pakistani Army is the premier security-providing agency in Balochistan and a key player in CPEC.

DOI: 10.4324/9781003413905-1

Many amongst the Baloch people consider the CPEC as economic colonisation and a threat to their very survival (Aamir, 2022; Zeb, 2019, p. 190) and since 2013 the ongoing fifth round of insurgency is mainly focused on sabotaging CPEC projects in the province of Balochistan (Adeel, 2015, p. 124; Sabharwal, 2022, p. 64; Wilkens, 2015). Unlike in the past, this time the Pakistani Army is not focused only on kinetic means to end the ongoing fifth round of insurgency in the province, but it has adopted more balanced approaches to counterinsurgency (COIN) as a consequence of institutional learning while fighting alongside US forces in the Global War on Terror (GWOT). A testament to the institutional COIN learning is the fact that 2013 was the very first time the Pakistani Army formally enacted its COIN doctrine despite fighting insurgencies for the last six decades. As a result, the COIN strategy of the Pakistani Army in Balochistan has been greatly transformed from the mere kinetic to inclusion of the critical components of the COIN like political primacy, winning hearts and minds, rules of engagements and certain other ethical and humanitarian aspects. All these renewed dimensions of the Pakistani COIN, their implementation on the ground, and their efficacy will be thoroughly discussed in the succeeding chapters of the book.

CPEC & Balochistan Insurgency

CPEC

South Asia is known as a region with conflict, political instability, and economic underdevelopment. It is argued that cooperation and development can bring stability to the region (Bhattacharjee, 2015, p. 1). CPEC is one such avenue; this project is of high economic impact which can strengthen the prospects of stability in Pakistan. During the state visit of Chinese Premier Li Keqiang to Pakistan in May 2013, a memorandum of understanding (MoU) about the long-term economic corridor was officially signed. However, this vision of CPEC found its eventual shape during the state visit of Chinese President Xi Jinping to Pakistan in April 2015 when he announced around $46 billion in development deals for the CPEC (Bhattacharjee, 2015, p. 2). This amount was increased to $62 billion by April 2017 (Garlick, 2018, p. 519).

Primarily, BRI, a vision announced by Xi Jinping in 2013, is aimed at reviving ancient trade routes connecting Asia with Africa and Europe (see Appendix 2). BRI, spearheaded by China, is a combination of two ambitious initiatives; the Silk Road Economic Belt (SREB) and the 21st Century Maritime Silk Route Economic Belt (CMSR) (Tiezzi, 2015; Ze, 2014). The SREB envisages the revival of the ancient Silk Route highways, through railroad and distribution networks, along with fibre optics, which once connected China to Europe by land. The SREB is planned to have three distinct corridors (Wolf et al., 2017, p. 100; Xinhua Silk Road, 2021). The Northern Corridor connects Beijing with Moscow and Germany. The Central Corridor connects the Chinese city of Shanghai to Europe via Iran. It passes through Tashkent, Tehran, and to Bandar Imam Khomeini Port of Iran on the Persian Gulf which stretches up towards Europe. This Iranian part of the Central Corridor, though a longer one, can serve as an alternative to CPEC in case CPEC does not materialise because of the overall security situation in Pakistan especially in the province of Balochistan or owing to the lack of management capacities (Wolf et al., 2017, p. 100). Lastly, the Southern Corridor starts from the city of Guangzhou (southern China) and moves to the city of Kashgar (Xinjiang Province) and into Pakistan at Khunjarab Pass. From Khunjarab in northern Pakistan, the Southern Corridor is linked to Gwadar Port (Balochistan) in the south, a gateway to the Arabian Sea, the Indian Ocean, and the Persian Gulf (Rana, 2015; Wolf et al., 2017, p. 100). The Southern Corridor from Khunjarab Pass to the Gwadar Port (Balochistan) will be implemented through the CPEC. The CMSR, a sea route, begins in the South China Sea passing through the Indian Ocean and South Pacific Ocean. It compliments SREB with the aim of connecting China to the Mediterranean Sea via the Persian Gulf (Wolf et al., 2017, p. 100). Both SREB and CMSR serve as the next stage in expanding China's commercial, strategic, and political influence ever since the Chinese "Go Global" policy was unleashed in 2000 (Bhattacharjee, 2015; Chung, 2018; Wang & Miao, 2016). The CPEC is located at the confluence of SREB and CMSR (see Appendix 2); therefore, it is a major project of the BRI initiative.

CPEC is a game-changer initiative of the South Asian Pacific Rim (Farooq, Rao & Shoaib, 2023). CPEC's importance has increased multi folds, and so has the project's security aspect considering Pakistan's present economic situation and energy crisis. The country is currently confronting a multifaceted crisis. Its economy is on the brink of collapse due to multiple factors including potential political turmoil, the rupee plummeting, historically high inflation rates, sagging growth, and a significant energy shortfall (Turak, 2023). The central bank has implemented a substantial increase in interest rates to counteract the effects of a weakened currency. Common citizens feel the tangible effects of rising food and fuel costs, and the economic difficulties are further compounded by the widespread destruction caused by recent flooding. Although these crises have led to the slowing pace of CPEC projects, it would be fallacious to presume that the China-Pakistan relationship is facing significant challenges (Kugelman, 2023).

Amidst all these crises, Pakistan and China have reaffirmed their commitment to CPEC and vowed to expand it further during the third round of Pakistan-China bilateral political consultations (BPC) held in Beijing on 18 March 2023. Pakistan sees foreign direct investment (FDI) as a major benefit of the CPEC from an economic perspective (Chang, 2014; United States Department of State, 2022). The Pakistani government hopes that the infrastructure development and energy improvements resulting from CPEC will help alleviate the current energy crisis[1] and have positive spillover effects on other economic sectors (Wolf et al., 2017, p. 103; Huaxia, 2022). By earmarking economic nodes along the route of CPEC in the industrialised urban centres and underdeveloped rural areas, Islamabad expects that it will help in providing an economic boost to the "poorer provinces" like Balochistan. This will ultimately result in reducing unemployment, poverty alleviation, and preventing the capital flight which has reached an alarming rate. It is also expected that the foreign reserves of the country will increase considerably as the CPEC progresses further.

Moreover, given the present geopolitical scenario, it is almost certain that China and Pakistan will continue their alliance. The escalating rivalry between the United States and China has imposed limitations on collaboration between Pakistan and the United States, thereby intensifying Islamabad's dependence on China for economic and military aid. Additionally, Pakistan's pursuit of a stronger relationship with Russia (Pakistan Today, 2023) comes amidst the deepening alliance between China and Russia.

Many security experts and researchers agree in their views that the CPEC is a strategic project for China. They believe it is unlikely that China will easily abandon CPEC, owing to the precarious economic condition of Pakistan or even if it comes under attack from various terror groups in the country. For instance, in April 2023, General Zhang Youxia, the vice chairman of China's Central Military Commission (CMC), remarked that China would work with Pakistan's military to "further deepen and expand" the two nations' mutual interests and jointly protect regional peace and stability. These remarks were made during a meeting between Zhang Youxia and the Pakistani COAS General Asim Munir on the latter's maiden official visit to China (Hussain, 2023).

In views of these security experts and researchers, the biggest interest China seeks from CPEC is an alternative route for her energy imports via the Strait of Malacca (see Appendix 3). Thus, in case of any potential blockade, China can easily bypass the Strait of Malacca and her trade, especially the oil supplies, will not be choked. That's how CPEC also provides a potential solution to China's Malacca dilemma (Lindley, 2022; Qadeer 2022, Interview, 08 December; BBC News, 2015; Chang, 2014; Hassan, 2020; Pillalamarri, 2015). This adds a strategic dimension for China in the CPEC project (Wolf et al., 2017, p. 104). Moreover, CPEC will shorten the route for China's energy imports from the Middle East by 12,000 kilometres (Bhattacharjee, 2015, p. 2; Hassan, 2020) and link it to the comparatively underdeveloped and landlocked western province of Xinjiang (Tharoor, 2015). This will help in the economic development of the underdeveloped western Chinese region in parity with the prosperous eastern parts of the country (Weihong, 2015; Ze, 2014). The western Chinese region is Beijing's top domestic security concern. Authorities in Beijing expect that by bringing economic development and stability to the Xinjiang province, they will be able to curtail the "*three evils*", namely *separatism, terrorism,* and *religious fundamentalism*[2] (Wolf et al., 2017, p. 104).

On the other side, the Chinese authorities have become more sceptical about the security threats in Pakistan (Kugelman, 2023). Several recent assaults have aimed explicitly at Chinese investments and citizens in Pakistan, such as the attack on a dental clinic in Karachi in September 2022, the Confucius Institute in Karachi in April 2022, and a luxury hotel in Quetta (Balochistan) that hosted a senior Chinese delegation in 2021. Baloch Liberation Army (BLA), a secessionist insurgent organisation based in Balochistan, claimed responsibility for these attacks except the last one on a luxury hotel in Quetta. Following Prime Minister Sharif's meeting with Chinese President Xi Jinping in Beijing in November 2022, the Chinese foreign ministry issued a statement stating that Xi had expressed his great concern about the safety of Chinese nationals and investment projects in Pakistan (Gul, 2022). All this insecurity to the Chinese investment projects of CPEC is mainly linked to the ongoing fifth round of insurgency in Balochistan.

The current round of Balochistan insurgency is the most violent round so far and is targeted towards the CPEC which promises prosperity and progress in Pakistan and has started a new era of proximity to China (Elahi, 2015; Shakil, 2022). The location of many of the projects of CPEC is Balochistan province. This corridor will incorporate 2,000 kilometres of transport links between Kashgar in northwestern China to Pakistan's Gwadar Port on the Arabian Sea via roads, railways, and pipelines. The insurgents' efforts after 2013 are mostly focused on sabotaging the CPEC by destroying infrastructure, killing the local/Chinese nationals working on the project and bombings in Balochistan.

Balochistan Insurgency

Pakistan has been in crisis since its inception with regard to the highly challenging problems of the management and incorporation of ethnonational identities in the country, especially in Balochistan (Samad, 2014, p. 293). Balochistan was given provincial status (the largest in terms of area and with the smallest population) after merging into the federation in August 1947. The province has been the site of sporadic armed revolts by the Baloch sub-nationalists since 1948, a year after the creation of Pakistan (Brown et al., 2012, p. 1; Hashmi, 2015, p. 68). The second, third, and fourth rounds of insurgencies were in 1958–59, 1963–69, and 1973–78 (Hasnat, 2011, p. 91), respectively. The current Balochistan insurgency started in 2006 and is the fifth round of insurgency (Adeel, 2015, p. 124; Sabharwal, 2022, p. 64; Wilkens, 2015). The ongoing fifth round of the Balochistan insurgency was triggered in 2006 when the Pakistani Military started fully-fledged armed action against the Baloch insurgents in the Kohlu District of Balochistan (Bansal, 2008, p. 182).

In August 2006, the province of Balochistan descended into the chaos of a new cycle of bombings and violence after the killing of Nawab Akbar Bugti, the chieftain of Bugti tribe, by the Pakistani Army during a military operation (Zeb, 2019, pp. 160–161). The Pakistani Army justified the extreme use of force against the Baloch insurgents in the name of the "War on Terror" to which President General Pervez Musharraf was already an ally to the United States. The Western allies of Pakistan disregard the issue as General Musharraf was following their tactics of targeted killing and renditions (Samad, 2014, pp. 294–295). At this point, the separatist demand by the Baloch took on a renewed zeal, isolated the Baloch political leadership from the Federal Government, further infuriated the Baloch and an insurgency in full swing started in the province (Samad, 2014, p. 295).

Baloch chronicles depict a constant picture of betrayal, deprivation, and treachery by the central government in Islamabad. The earlier rounds of the insurgency, though successfully crushed with heavy kinetic actions, left permanent marks of distrust between the Baloch nationalists and the central government. As Catignani highlights: "Clearly, a military can (poorly) manage, but not resolve an insurgency through military means alone" (2012, p. 273). This is the case with these earlier rounds of the Balochistan insurgency. The Pakistani Military could suppress the earlier rounds by excessive use of force only, while completely relegating the other essential elements of the COIN like political aim, political primacy, etc., but it could not bring an end to the insurgency. Therefore, the insurgency kept on resurfacing in the Balochistan province. Each round of the Baloch insurgency was more intense and organised than the former and gathered even more popular support amongst the Baloch tribes (Bansal, 2008, p. 184). In 1973, 80,000 Pakistani troops fought a COIN campaign against 55,000 Baloch insurgents. The Pakistani troops were operationally not only supported by the fighter jets of the Pakistan Air Force but also by the Iranian Air Force (Bansal, 2008, p. 184). It is estimated, though death toll estimates vary, that between 7,300 and 9,000 Baloch insurgents and civilians were killed while the Pakistani Military lost 300–400 soldiers during the fourth round of insurgency (Paul et al., 2013b, p. 361). The Pakistani Military used force brutally to suppress the fourth round of Balochistan insurgency at all costs after their defeat in Bangladesh in 1971 (Bansal, 2008, p. 184). The Baloch insurgents were anticipating a Soviet intervention, but somehow it did not materialise (Bansal, 2005, p. 252). As a consequence of the extreme use of force against the Baloch, the then President and the Chief of the Army Staff of Pakistan General Zia ul Haq granted amnesty to the insurgents (Amin, 1998, pp. 176–177; Zeb, 2019, p. 130). Thereafter, the Baloch nationalists were always part of the Provincial Government in Quetta (Balochistan).

Contemporary Conflict Drivers

The current round of Balochistan insurgency was rekindled by a combination of factors including long unaddressed historical grievances about the political and economic suppression of the Baloch people, the construction of the Gwadar mega-port, poor management of differences leading to an unjustified distribution and control of resources (including natural resources like natural gas and minerals), the war in neighbouring Afghanistan, and the Pakistani Military's harsh attitude towards the Baloch nationalist demands (Kupecz, 2012). These conflict drivers, some of which are rooted deep into the historical legacy of suppression and subjugation, have made the conflict very complex and its likely resolution improbable. The Baloch grievances can be broadly classified under political, socio-economic, and security domains, as shown below in Figure 1.1.

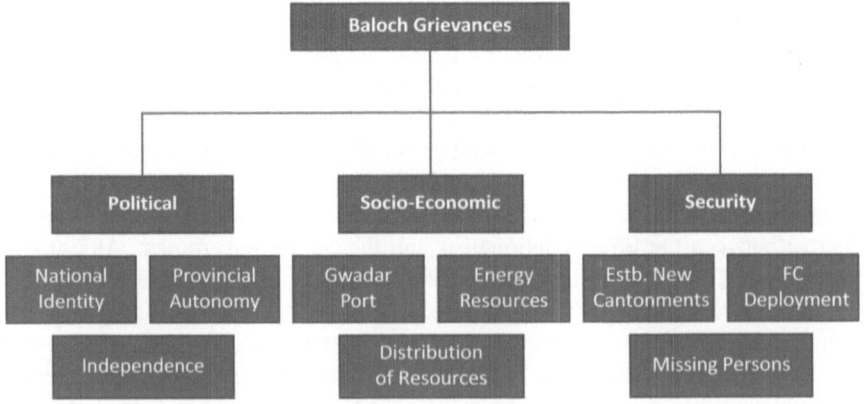

Figure 1.1 Baloch Grievances.

Source: Author

Political Grievances

National Identity

National identity and provincial autonomy are the two political demands which the Baloch nationalists have been pursuing since Balochistan became part of Pakistan in the late 1940s. At the time of the partition of the Indian subcontinent, Balochistan was comprised of four princely states, namely Makran, Las Bela, Kharan, and Kalat. All of the above mentioned princely states, except Kalat, joined the Federation of Pakistan in 1947 (Mahsood & Miankhel, 2013, p. 52). The Khan of Kalat Mir Ahmad Yar Khan declared the independence of Kalat as per the option available in the Partition Plan of India-1947 (Bennett-Jones, 2009, pp. 132–133). In Kalat, both houses of parliament unanimously refused the accession of the state with Pakistan (Priyashree, 2006, p. 3). In the aftermath of this decision, on 1st March 1948, the Pakistani Army entered Kalat as a result of the failure of negotiations for the peaceful settlement of the issue and forced the Khan of Kalat to accede to Pakistan (Bennett-Jones, 2009, p. 133). Thus, Mir Ahmad Yar Khan signed the instrument of accession on 27 March 1948. As a result, Prince Abdul Karim, the brother of the Khan of Kalat, revolted in April 1948, starting the first-ever Baloch insurgency after the partition of India in 1947 (Axmann, 2012; Bangash, 2015).

The second Baloch insurgency surfaced in 1958 after the government of Pakistan promulgated the "One Unit Plan" (Mahsood & Miankhel, 2013, p. 69) which led to the suppression of Baloch identity and rights. As per the "One Unit Plan", Balochistan, Punjab, Sindh, and North West Frontier Province (now KPK) were amalgamated into one administrative unit.

Again in 1973, the federation attempted to threaten the Baloch identity and rights by dissolving the elected Provincial Government of Balochistan by the then Prime Minister of Pakistan, Zulfiqar Ali Bhutto on charges of treason against the federation. Bhutto established the governor's rule in the province of Balochistan which facilitated the direct control of the central government over the province through the governor, a representative of the federal government (Mahsood & Miankhel, 2013, p. 52). This stimulated another round of insurgency (1973–78). This round of insurgency ended in 1978 after the imposition of Martial Law and removal of Bhutto's government by General Zia ul Haq in July 1977 and subsequently granting amnesty to the Baloch insurgents in 1978.

Provincial Autonomy

The core issues of provincial autonomy for Balochistan, Baloch people's rights and the question of their ethnic identity are interwoven (Hashmi, 2015, p. 64). Since 1948, the Baloch leaders have been raising their voices for maximum autonomy within the federation and protection of their ethnic identity by not altering the boundaries of the province, but by using the commitments made with the Baloch people at the time of their merger with Pakistan. The dilemma is that none of the civilian governments or military regimes was ready to give provincial autonomy. Rather, they all focused on the centralisation of power with the only exception of the Pakistan People's Party (PPP) government (2008–13). The PPP's government introduced special reforms for the provincial autonomy of the province (known as the Balochistan Package), passed the 18th amendment of the constitution, which abolished the concurrent legislative list from the 1973 constitution of Pakistan, a hurdle in provincial autonomy (because the centre could override provincial legislation on all subjects listed on it) and thus a subject of considerable tension between the centre and the provinces (Adeney, 2012, p. 547). The 18th Amendment went one step further and provided that the provinces shall be given a greater share in the distribution of resources between the centre and the provinces[3] (Adeney, 2012, p. 548). The changes to the distribution of resources were important, for symbolic as well as financial reasons (Adeney, 2012, p. 548). The PPP government was unable to fully implement these reforms for addressing the grievances of the Baloch people because the ever-strong military establishment in Rawalpindi was not happy on the issue of greater provincial autonomy for Balochistan. The government of Balochistan was not satisfied with the PPP government's political efforts for provincial autonomy (Fair, 2012, pp. 28–30).

Socio-Economic Grievances

The socio-economic grievances of the Baloch people have always revolved around Gwadar Port (even well before the CPEC), control of energy resources in the province and the distribution of resources. The late Akbar Bugti, a political figure who also served as the chief minister and governor of Balochistan province, had severe differences with the central government in Islamabad on issues like the construction of Gwadar Port and the terms and conditions for natural resource extraction, particularly natural gas, from Balochistan. The Baloch nationalist leaders demanded the fair share of the revenues from natural resources in parity with the other provinces (Priyashree, 2006, p. 5). Likewise, the Baloch nationalist leaders always complained about the development of the Gwadar seaport without consultation with the Balochistan government and argued it was not to the benefit of the Baloch people. The Provincial Government has no control over Gwadar (Mahsood & Miankhel, 2013, p. 54). This helped to bring back the age-old grievances of Baloch nationalists against the Federal Government. As a result of this situation, a low-intensity insurgency in Balochistan was triggered which later took full momentum after the killing of Akbar Bugti and is still ongoing. The triggering of insurgency raised serious concerns in Beijing about the security of Gwadar and hence the possibility of investing in Gwadar Port (Boni, 2016, p. 503).

Security Related Grievances – Enhanced Army Influx and Forced Disappearances of Youth

The simmering tension between Baloch nationalist tribal chiefs and the government was aggravated after General Musharaf, in 2005, announced the increased presence of the regular army in Balochistan for the development and protection of mega projects like Gwadar. At that time Musharraf had invested much of his political credibility in megastructure and infrastructure development for reviving the weak economy of the country with the Gwadar Port as the mainstay of his economic uplift strategy (Boni, 2016, p. 504). Earlier, the Baloch nationalists greatly opposed the Federal Government's decision to allow the United States to establish its bases in Balochistan at Gwadar, Pasni, Jacobabad, and Dalbandin as part of the collaboration in the GWOT during General Musharaf's regime (Hashmi, 2015, p. 74). Furthermore, the announcement of the establishment of new military cantonments in Balochistan at Khuzdar, Gwadar, Dera Bugti, and Kohlu during General Musharaf's regime on the pretext of enhanced security also met with strong resistance from the Baloch nationalists.

At the outbreak of a full-scale insurgency (the fifth round) after the killing of Akbar Bugti in 2006, the forced disappearances of people in Balochistan on the suspicion of being insurgents or collaborators/facilitators of insurgents and subsequently the extrajudicial killings of a few in captivity by the Frontier Corps (FC) (which is a paramilitary organisation in Balochistan), Inter-Services Intelligence Agency (ISI) and the Army created a lot of unrest and fear amongst the people of Balochistan. As per the records of Voice of Baloch Missing Persons (VBMP), a human rights organisation that gathers data and collects information of Baloch missing persons, there were more than 2,825 documented cases of enforced disappearances of Baloch activists from 2005–2014 (Hashim, 2014). The practice of forced disappearances, especially of Baloch youth, continued by the law enforcing agencies and military as late as October 2018. The issue of missing persons will be thoroughly analysed in Chapter 6.

Research Puzzle and Research Purpose

The extent to which the Pakistani Army's Counterinsurgency (COIN) strategy (2013–2022) is effective in the eradication of perceived threats for the CPEC in Balochistan is the research puzzle of this book. Therefore, the book provides an overall assessment of the effectiveness of the Pakistani Army's conduct of COIN warfare in Balochistan after conceptualising COIN. Additionally, the correlation between the flexibility of COIN strategies and the likely outcome of COIN campaigns in Balochistan by the Pakistani Army is explored by the study. It further analyses the limit of the flexibility of the COIN strategy and observes what must remain unchanged. Furthermore, the study determines how rigid or flexible the adopted COIN strategies were, considering the nature of changes introduced to the strategies during the conduct of COIN warfare and the outcomes that can be traced. As such, the research provides a holistic assessment of the results of the COIN campaign in Balochistan and compares the flexibility of the strategies with their effectiveness. This research aims to answer the primary research question: *What is the effect of the Pakistani Army's COIN Strategy (2013–2022) in reducing the perceived threats to the China-Pakistan Economic Corridor (CPEC) in Balochistan?*

This research traces the evolution of the Pakistani COIN doctrine and analyses the efficacy of the COIN strategy adopted by the Pakistani Military in Balochistan. The study seeks to understand the difference and equilibrium between

COIN strategy and its practical manifestation on the ground after formally enacting the doctrine. Thus, the time period has been distinctly divided into two, i.e., from 2006–2012 (prior to the COIN doctrine) and 2013–2022 (after formally enacting COIN doctrine). This leads to the sub-research question: *"What is the Pakistani Army's Doctrinal Approach to Counterinsurgency?"*

Along these lines, the purpose of the study is to characterise the Pakistani Army's role, assistance, approach, and the campaigns in Balochistan using Kilcullen's (2006) *"three pillars"* model of COIN. The three pillars are classified as Security, Political, and Economic. This study examines how the principles under each pillar were realised in practice by the Pakistani COIN forces focusing precisely on understanding which principles were fully realised, which were altered, which were abandoned altogether and how the strategies were adjusted to the operational environment. Furthermore, it identifies and analyses the weaknesses of the COIN strategy in Balochistan and finally proffers some policy recommendations, based on the framework for whole-of-government COIN. The book analyses the COIN strategy being employed by the Pakistani Security Agencies in the province of Balochistan for creating a safe and secure environment for the CPEC. The current insurgency in Balochistan has been caused by both internal and external factors which pose a big challenge for the Pakistani Security Agencies to counter in order to achieve a favourable environment for smooth development and functioning of the CPEC project (Iqbal, 2018). Moreover, the presence of the Islamic State of Khorasan (ISKP)[4] in Balochistan and its alliances with the Baloch secessionist organisations and the sectarian militant organisations in the province has made the security situation more complicated and at the same time given an international outlook to a regional conflict (Ashraf, 2017; Jadoon, 2018). This leads to another sub-research question which asks: *"How is CPEC in Balochistan under threat?"* Thus, the study aims to address the questions concerning the utility of a COIN strategy for creating a safe environment for CPEC, the efficacy level of the COIN strategy employed in the context of CPEC in Balochistan and additional measures to enhance the existing COIN strategy's viability for smooth trade operations through CPEC.

Contemporary discourse on the adoption of correct COIN strategy, which gives the best chance of the government prevailing is vast and often contentious (Paul et al., 2013a, p. xviii). A host of concepts and areas of emphasis are advocated, but more often such advocacy is based on relatively limited evidence (Paul et al., 2013a, p. xviii). There is no published empirical research on the utility of COIN as an operational tool in the context of CPEC and Balochistan. Hence, this research broadly addresses the concept of evaluating the utility of COIN practices by the military for achieving a specific aim in a specific area and in a specific time frame with CPEC as a model. The book will thoroughly discuss the adoption of a COIN strategy by Pakistani Security Forces against the ongoing insurgency in Balochistan. The analysis will be based on the insurgent approaches and responding COIN practices. Various COIN concepts ranging from "classical era" concepts like *Hearts & Minds, Pacification, Reform* and *Redress, Unity of Effort, Crush Them* to some contemporary concepts for COIN like *Strategic Communication/ Information Operations, Clear, Hold, and Build; Boots on the Ground, Criticality of Intelligence* and *Tangible Support Reduction* as and when applied by the Security Forces in Balochistan will be analysed, and their efficacy will be tested for classifying them into "Good or Bad COIN practices" (Paul et al., 2013a, p. 249). Overall, this book is an attempt to understand and rationalise the apparently paradoxical strength and weakness of the Pakistani state machinery in COIN movement in Balochistan against the backdrop of CPEC. In the next chapter, the research methodology and conceptual framework will be comprehensively discussed in the context of academic discourse.

Existing Void about Pakistani COIN

Pakistan has been facing the menace of fully-fledged insurgencies in different regions of the country since 1948. The Pakistani Army has conducted various COIN operations in the varied ecological landscapes (Gates & Roy, 2014). The three most important regions where the Pakistani Army has conducted COIN, in chronological order, are East Pakistan (now Bangladesh), Balochistan, and the North West Region (comprising FATA and the province of KPK, earlier known as the North West Frontier Province [NWFP]). Currently, the Pakistani Army is fighting an insurgency in Balochistan (Gates & Roy, 2014). A very limited number of studies are available on the Pakistani Army's COIN efforts (Nawaz, 2011, p. 3; Gates & Roy, 2014, p. 4).

Reasons for Non-Availability of Research on Pakistani COIN

The major reason for the acute scarcity of literature on Pakistani COIN in the public domain is the strict policy of the Pakistani Defence Forces of putting an "iron curtain", in the name of secrecy, on the operational and administrative details comprising documents and oral history of various campaigns including both conventional and sub-conventional carried out by the Pakistani Army since its inception. Owing to this strict iron curtain policy, no access is granted to any writer, academic, researcher or journalist (except for a couple of exceptions) to information about the Pakistani Army's campaigns in general; even decades-old archival material related to various military campaigns is inaccessible despite not being politically and militarily sensitive. This phenomenon seems to be common to the post-colonial armies of the subcontinent. Gates and Roy (2014, p. 2) highlight that even the archival documents for the post-1947 period of the Indian Army are not open to public scrutiny and moreover, the senior officers are "tight-lipped" about any issue of national security including various military campaigns. At times, some of them agree to verbally confirm limited amounts of information but strictly on the condition of anonymity; the researchers suggest that "such an attitude makes a mockery of the empirical value of their evidence" (Gates & Roy, 2014, p. 2). The same attitude is observed in the case of the Pakistani Army's actions in general and COIN campaigns in particular[5].

This extremely strict secrecy policy leads to the inaccessibility of information and ultimately results in acute scarcity of the available raw material comprising of facts and figures as well as published literature on Pakistani COIN. A reason for this scarcity can be traced to the British colonial legacy of the Indian Official Secrets Act of 1923 – a faithful replica of the British Official Secrets Act 1911. Although Britain reformed the law in 1989 (Noorani, 2015), the Indian Official Secrets Act of 1923 is still on Pakistan's statute books, like India. Indian Official Secrets Act is a part of the Pakistan Army Act (PAA) in its originality except for the omission of the word *Indian*. Now it is called the Official Secret Act (OSA) of 1923, an anachronistic law and a legacy of corrupt colonial practices.

In the Pakistani Army, the OSA of 1923 is interpreted rigorously to mean the non-disclosure of any details about the army pertaining to any operational or administrative issue whether "useful to an enemy" or not, no matter how dated that information is. This practice has been strictly followed, particularly under military regimes. To comply with the OSA of 1923, all officers and troops are made to read all the clauses of the Act once a month throughout their active service and they have to officially render a written certificate with signatures to that effect. The OSA 1923 forms an important part of the curricula of the basic military training in the defence force academies during various professional courses and exams throughout the military service. Therefore, not sharing anything with anybody outside the military circle pertaining to any aspect of the military operations, decision-making process, administrations, logistics, tactics and strategy, etc., becomes engrained in all the uniformed personnel of the Pakistani Army, irrespective of their rank. Even within military circles, information sharing is practised only on a "need-to-know basis". This explains why there is no COIN literature, based on primary sources, available about the Pakistani Army[6].

After the military regime of General Musharraf, the Pakistani Army during General Ashfaq Kayani's tenure (2007–13) gradually adopted a more open and flexible approach towards releasing details about its various COIN operations and strategies. In January 2013, the then Director General (DG) of the Inter Services Public Relations (ISPR), Major General Asim Saleem Bajwa[7], in a press briefing released details about the first-ever conceived Sub-Conventional Warfare (SCW) Doctrine of the Pakistani Army (The Express Tribune, 2013). This was an unprecedented move since there has been a more open and flexible approach.

Unlike the British, United States and other modern armies, there is no trend of retired military officials of the Pakistani Army writing about their war experiences in the form of a book or a journal article. A few pieces are available, but they date back to the 1965 and 1971 Indo-Pak wars and are mostly historical accounts only, not focusing on the critical evaluation of the war strategy[8]. No retired officer has ever written about the COIN operations of the Pakistani Army. This also explains why the literature on COIN is silent in Pakistan.

Moreover, another factor which contributes to the non-availability of the literature on Pakistani COIN is the practice of media censorship/blackout by the Pakistani Army in the areas where it conducts the COIN operations. This media censorship/blackout continues long after the operations which means that no academic or journalistic literature can be produced on the subject. For instance, during the Pakistani Army's COIN campaign in East Pakistan, the Pakistani Military confined all foreign reporters and journalists to the Hotel Intercontinental in Dhaka on March 25, 1971 – a night prior to the launch of the COIN campaign by the Pakistani Army in that area. Two days later, as Dhaka burned, the reporters were expelled from the country – their notes and tapes were confiscated (Mostofa, 2017). The media censorship/blackout is still in practice in Balochistan as a matter of unwritten policy (Freedom Network, 2017; Khattak, 2018; Nadir, 2022; Nasar, 2016). Therefore, practically, it is nearly impossible for academics and journalists to write on the topic. Overall, there is not much-published literature available on the Balochistan security situation, and its connection to the CPEC and there is a large void on the current round of insurgency in Balochistan and corresponding COIN strategy of the Pakistani Army.

Limited research is available on the COIN strategy of the Pakistani Army with specific reference to FATA insurgency. The insurgency in the tribal region of Pakistan (FATA) bordering Afghanistan and the government responses, including the Pakistani Army's COIN operations there, is mainly studied as a result of US-NATO involvement in Afghanistan in the aftermath of Operation Enduring Freedom but there are a few articles on the subject (Gates & Roy, 2014, p. 4, 127). The few comprehensive and famous pieces of research on the subject covering both the insurgency and to some extent the corresponding COIN strategy of the Pakistani Army were written by Fair and Jones (2009, 2010) and Nawaz (2011). These pieces of literature are extensively quoted in any study on the Pakistani Army and insurgency in FATA. The detailed information was provided to these US-based and affiliated researchers by the Military Operations (MO) Directorate of the Pakistani Army during an operational briefing in 2009/10 in an unprecedented move with a political agenda behind it.

Therefore, this book is the first and original effort to comprehensively analyse the ongoing fifth round of Balochistan insurgency and the efficacy of the corresponding COIN strategy by the Pakistani Army against the backdrop of the security of CPEC in Balochistan. Thus, the book fills the existing research void on the subject.

Organisation of the Book

The book is structured into seven chapters, through which the analysis is articulated. After the Introduction, Chapter 2 delves into the book's conceptual framework and research methodology. As previously stated, the book assesses the efficacy level of the COIN strategy employed in the context of CPEC in Balochistan by the Pakistani Military using David Kilcullen's (2006) "three pillars of COIN model (3PCM)" as the conceptual framework. The chapter elaborates on the conceptual framework and its two parts: the "conflict ecosystem", and the second is a framework for the whole-of-government COIN in that environment. Moreover, the suitability of the framework for the current study will be comprehensively discussed in the context of academic discourse. The COIN scorecard (attached as Appendix 6) developed by Paul et al. (2013a, p. 249) will also be discussed in detail to complement Kilcullen's 3PCM as a yardstick for characterising the Pakistani Army counterinsurgency practices in Balochistan (after, 2013) into "Good" and "Bad". Finally, the chapter elaborates on numerous methodological tools for harnessing empirical information and acquiring relevant modus operandi for presenting the research findings.

Chapter 3 establishes what doctrine is and sets the development of the Pakistani Army's COIN doctrine in a historical context. It traces the Pakistani Army's culture of fighting *small wars* which it inherited in 1947 from the British Indian Army. Then it looks at the step-by-step development of the institutional learning of the Pakistani Army towards fighting insurgency, using Downie's Institutional Learning Cycle, also highlighting the *Lessons Learned* during various COIN campaigns/ periods which depict the Pakistani Army's flexibility and willingness to learn. This enables us to trace the evolution of Pakistani COIN doctrine, and the international and domestic context/environment in which it was developed.

Chapter 4 analyses the institutionalisation of the learning of the Pakistani Army concerning COIN, as thoroughly discussed in Chapter 3, since independence in 1947. It explains the institutionalisation process of the learning, initiated by General Kiyani (former Chief of the Army Staff) and the efforts and stimuli to reach the consensus of promulgating the SCW doctrine. This is followed by tracing the change in the *organisational behaviour* of the Pakistani Army through the training regimes. Thereafter, the chapter looks at the themes enshrined in the SCW doctrine and how it addresses the performance gaps of the army highlighted in Chapter 3.

Chapters 3 and 4 set a firm foundation for the later empirical chapters of the book to evaluate the extent to which the Pakistani Army adheres to the doctrinal approach enshrined in the SCW document (2013) by incorporating it into the COIN strategy in Balochistan. These later chapters will also examine the efficacy of the Pakistani Army's COIN strategy in Balochistan after 2013, particularly in relation to the safety of CPEC.

Chapter 5 deals with the threat matrix to the CPEC. It highlights various geopolitical challenges concerning the security environment in Balochistan. It will focus on analysing and evaluating various external threats vis-à-vis, the prevailing geopolitical security situation in Balochistan, and the adjoining area of Iran and Afghanistan. The Indian interest of influencing her sphere of interest in the region and gaining access to the Central Asian Energy corridor will also be discussed as it is of significant importance to the security of CPEC.

Apart from external threats, there are a diverse set of internal threats for CPEC which will be explained in detail in the second part of the chapter. The biggest internal threat for CPEC is the ongoing secessionist insurgency in Balochistan followed by the activities of sectarian militant organisations in the province. Other threats include the converging ideologies and interests of the Islamic State of Khorasan (ISKP) in Balochistan with the secessionist insurgent organisations and the sectarian militant organisations. This scenario increases the hostile environment for CPEC in Balochistan.

Chapter 6 deals with characterising the Pakistani Army's role, assistance, approach, and the campaigns in Balochistan using Kilcullen's three pillars of COIN model (3PCM) from 2006–2012 (prior to the COIN doctrine) and 2013–2022 (after formally enacting the COIN doctrine). The continuity of each pillar and the balance between them as maintained by the Pakistani Army is discussed in detail. The role of the military is also highlighted in-depth, which apparently seems part of the security pillar only, but of course, military means are applied across the model, not just in the security domain. Chapter 6 evaluates two aspects. Firstly, the extent to which the Pakistani Army adheres to the doctrinal approach, as enshrined in the SCW document (2013), by incorporating it into the COIN strategy in Balochistan. Secondly, the efficacy of the COIN strategy of the Pakistani Army in fighting insurgency in Balochistan after 2013 concerning the safety of CPEC. Diverse data in the form of several graphs/charts have been used as empirical evidence. These data show the various yearly trends, such as fatalities as the result of state-based violence, the number of insurgents surrendered and bombing incidents before and after 2013 (the promulgation year of SCW doctrine and signing of a MoU for the CPEC). These data add credibility to the arguments and findings of the study.

In the last chapter, the empirical findings are summarised, compared, and contrasted. On the basis of these empirical findings, the conclusions are drawn. In addition, the chapter highlights the weaknesses of the existing COIN strategy in Balochistan and thus proffers some recommendations for renewed COIN policy-making in the local and global context for China/Pakistan in achieving their strategic goals of BRI/CPEC.

Notes

1 In February 2023, the Pakistani Prime Minister, Shehbaz Sharif, inaugurated a new Chinese-designed reactor at Karachi Nuclear Power Plant. China financed the $2.7 billion project to refurbish the facility and strengthen Pakistan's energy security (Mangi & Haider, 2023).

2 Although it is important to note that China is also using extensive coercive measures to achieve this result (Ramzy & Buckley, 2019; UN Human Rights Office (OHCHR), 2022).

3 Before the passing of the 18th Amendment, all "lands, minerals and other things of value within the continental shelf. . . were [vested] in the Federal Government". Revised Article 172 (as the result of 18th Amendment) provided that "mineral oil and natural gas within the Province or the territorial waters adjacent thereto shall vest jointly and equally in that Province and the Federal Government" (Adeney, 2012, p. 549). The revised Article substantially increases the resource base of provinces full of natural resources, such as Balochistan. That's how the 18th Amendment also sought to redress the concerns of Balochistan (Adeney, 2012, p. 549).

4 In May 2019, the Islamic State fragmented the Islamic State Khorasan Province (ISKP) into separate branches for Pakistan, Afghanistan, and India, resulting in the creation of the Islamic State of Pakistan (ISPP) (Webber & Valle, 2022). In July 2021, ISPP Wali Abu Mahmood announced through a statement that the Islamic State (IS) had transferred Khyber Pakhtunkhwa (KPK) Province in Pakistan to the ISKP administration, prompting the group to claim responsibility for any future attacks in the area under the name of ISKP (Webber & Valle, 2022). However, Durrani (2022, Interview, 11 August), a police officer serving in the Counter Terrorism Department (CTD) of Balochistan Police, claims that there is still ambiguity regarding the Islamic State's brand name in Balochistan province. While sometimes the organisation claims responsibility for terrorist attacks in Balochistan as ISPP, other times it is claimed as ISKP. Therefore, for clarity, the Islamic State in Balochistan will be referred to as ISKP throughout this book.

5 Rahimuddin Khan (late) [famously known as *"Rahim, the Brave & the Bold"* amongst his peers in the army] was one exceptional senior officer (four-star) who spoke with the author wholeheartedly and bluntly about the events and aftermath of the East Pakistan Insurgency of 1971 and the fourth round of Balochistan insurgency (1973–78). He was an eyewitness to these events as a senior army officer and later the Governor of Balochistan. His insights about these two insurgencies will be frequently quoted in Chapter 3. He was

also the only exception who did not put the condition of anonymity to quote his references in this research. Rahimuddin Khan was the nephew of famous educationist and politician Dr. Zakir Husain, who was the President of India from 1967 to 1969.

6 An interesting fact is that a number of so-called "restricted/classified" materials including the staff publications, training manuals, and study period presentations (MS PowerPoint slides/Word scripts) of the Pakistani Army are available on the internet; thanks to the advancement in digital technology and its role in information sharing. Such stuff is available on academic material websites/online digital library like Scribd®, Academia. edu®, and SlideShare®, etc. The same is the case with the Indian Army's training and operational publications. It is worth highlighting that all the documents about the Pakistani Army referred to in the book are accessed through the open web sources.

7 After relinquished as DG ISPR, Major General Asim Saleem Bajwa was promoted to the rank of Lieutenant General and posted as Commander Southern Command, Quetta (Balochistan) in September 2017. After retirement from the Army, in November 2019 Bajwa was appointed by the Federal Government as the Chairman of the CPEC Authority, established by transforming the CPEC Secretariat into an autonomous body under the Ministry of Planning, Development and Reform (Dawn, 2019).

8 See for example, Salik, S. (1977). *Witness to Surrender*. Karachi: Oxford University Press; Matinuddin, K. (1994). *Tragedy of errors: East Pakistan crisis, 1968–1971*. Lahore: Wajidalis Publishers; Siddiqi, A.R. (2005). *East Pakistan: The Endgame: An Onlooker's Journal 1969–1971*. Karachi: Oxford University Press; Amin, A.H. (2013). *Why Indian Army and Pakistan Army Failed in 1965 War.* Createspace Independent Publishing.

References

Aamir, A. (2022) 'China wants own security company to protect assets in Pakistan', *Nikkei Asia*, 28 June. Available at: https://asia.nikkei.com/Politics/International-relations/China-wants-own-security-company-to-protect-assets-in-Pakistan (Accessed: 16 April 2023).

Adeel, K. (2015). 'Renewed Ethno Nationalist Insurgency in Balochistan, Pakistan: The Militarized State and Continuing Economic Deprivation', in Chima, J. S. (ed.) *Ethnic Sub Nationalist Insurgencies in South Asia: Identities, Interests and Challenges to State Authority*. Abingdon, Oxon: Routledge, pp. 124–142.

Adeney, K. (2012). 'A Step Towards Inclusive Federalism in Pakistan? The Politics of the 18th Amendment', *Publius. The Journal of Federalism*, 42 (4), (Fall 2012), pp. 539–565. doi: 10.1093/publius/pjr055.

Amin, T. (1998). *Ethno-National Movements in Pakistan: Domestic and International Factors*. Islamabad: Institute of Policy Studies.

Ashraf, S. (2017). *ISIS Khorasan: Presence and Potential in the Afghanistan-Pakistan Region. Centre for the Response to Radicalisation and Terrorism*, The Henry Jackson Society, London. Available at: http://henryjacksonsociety.org/wp-content/uploads/2017/10/HJS-ISIS-Khorasan-Report.pdf (Accessed: 27 July 2022).

Axmann, M. (2012). *Back to the Future: The Khanate of Kalat and the Genesis of Baluch Nationalism, 1915–1955*. Oxford: Oxford University Press.

Bangash, Y. K. (2015). *A Princely Affair: The Accession and Integration of the Princely States of Pakistan, 1947–1955*. Karachi: Oxford University Press.

Bansal, A. (2005). 'The Revival of Insurgency in Balochistan', *Strategic Analysis*, 29 (2) (April–June), pp. 250–268. doi: 10.1080/09700161.2005.12049805.

Bansal, A. (2008). 'Factors Leading to Insurgency in Balochistan', *Small Wars & Insurgencies*, 19 (2), pp. 182–200. doi: 10.1080/09592310802061356.

BBC News (2015) *Is China-Pakistan 'Silk Road' a Game-Changer?* Available at: http://www.bbc.co.uk/news/world-asia-32400091 (Accessed: 15 August 2022).

Bennett-Jones, O. (2009). *Pakistan: Eye of the Storm* (3rd ed.). New Haven, CT: Yale University Press.

Bhattacharjee, D. (2015) 'China Pakistan Economic Corridor (CPEC)', *Indian Council of World Affairs*, 12 May. Available at: https://www.icwa.in/show_content.php?lang=1&level= 3&ls_id=5103&lid=835 (Accessed: 19 March 2022).

Boni, F. (2016). 'Civil-Military Relations in Pakistan: A Case Study of Sino-Pakistani Relations and the Port of Gwadar', *Commonwealth & Comparative Politics*, 54 (4), pp. 498–517. doi: 10.1080/14662043.2016.1231665.

Brown, M., Dawod, M., Irantalab, A., Naqi, M. & Carment, D. (2012) 'Balochistan Case Study-INAF 5493-S: Ethnic Conflict: Causes, Consequences and Management'. Available at: http://citeseerx.ist.psu.edu/viewdoc/download?doi=10.1.1.349.3925&rep=rep1&type= pdf (Accessed: 4 March 2022).

Catignani, S. (2012). 'Israeli Counterinsurgency', in Rich, P. B. & Duyvesteyn, I. (eds.) *The Routledge Handbook of Insurgency and Counterinsurgency*. Abingdon, Oxon: Routledge, pp. 263–275.

Chang, G. G. (2014) 'China's Big Plans for Pakistan', *The National Interest,* 10 December. Available at: http://nationalinterest.org/feature/chinas-big-plans-pakistan-11827 (Accessed: 11 March 2022).

Chung, C. P. (2018). 'What Are the Strategic and Economic Implications for South Asia of China's Maritime Silk Road Initiative?', *The Pacific Review*, 31 (3), pp. 315–332. doi: 10.1080/09512748.2017.1375000

Dawn (2019) *Asim Bajwa made chairman of newly created CPEC Authority*. Available at: https://www.dawn.com/news/1519047 (Accessed: 27 August 2022).

Elahi, N. (2015) *China-Pakistan Economic Corridor, Security Threats & Solutions: A Strategy*. Pakistan: Pakistan-China Institute. Available at https://www.pakistan-china.com/mn-monograph-2.php (Accessed: 24 February 2022).

Fair, C. C. (2012) 'Beyond Hearing Range', *Herald*, March. Karachi, Pakistan, pp. 28–29.

Fair, C. C. & Jones, S. G. (2009). 'Pakistan's War Within', *Survival*, 51 (6), pp. 161–188. doi: 10.1080/00396330903465204.

Fair, C. C. & Jones, S. G. (2010). *Counterinsurgency in Pakistan*. Santa Monica, CA: RAND Corporation.

Farooq, M., Rao, Z. R. & Shoaib, M. (2023). 'Analyzing the Determinants of Sustainability of China Pakistan Economic Corridor (CPEC) Projects: An Interpretive Structural Modelling (ISM) Approach', *Environmental Science and Pollution Research*, 30, pp. 12385–12401. doi: 10.1007/s11356-022-22813-3

Freedom Network (2017) *Special Report: Blackout in Balochistan: Media Reporting in Fear, Living Under Threat*. Available at http://www.fnpk.org/special-report-blackout-in-balochistan-media-reporting-in-fear-living-under-threat/ (Accessed: 08 October 2022).

Garlick, J. (2018). 'Deconstructing the China–Pakistan Economic Corridor: Pipe Dreams versus Geopolitical Realities', *Journal of Contemporary China*, 27 (112), pp. 519–533. doi: 10.1080/10670564.2018.1433483.

Gates, S. & Roy, K. (2014). *Unconventional Warfare in South Asia: Shadow Warriors and Counterinsurgency*. Abingdon, Oxon: Routledge.

Gul, A. (2022) 'China Urges Pakistan to Ensure Security of Chinese Working on Bilateral Projects', *Voice of America (VOA)*, 02 November. Available at: https://www.voanews.com/a/ china-urges-pakistan-to-ensure-security-of-chinese-working-on-bilateral-projects/6817008. html (Accessed: 8 December 2022).

Hashim, A. (2014) 'Families of missing Baluch march for justice', *Aljazeera*, 28 February. Available at: https://www.aljazeera.com/indepth/features/2014/02/families-missing-baloch-march-justice-2014227101949360898.html (Accessed: 23 July 2022).

Hashmi, R. S. (2015) 'Baloch Ethnicity: An Analysis of the Issue and Conflict with State'. *Journal of the Research Society of Pakistan*, 52 (1) (January–June), pp. 57–84. Available at: http://pu.edu.pk/images/journal/history/PDF-FILES/4-%20PC%20Dr.%20Rehana%20Saeed%20Hashmi_52-1-15.pdf (Accessed: 22 June 2022).

Hasnat, S. F. (2011). *Global Security Watch Pakistan*. Sanat Barbra, CA: Prager.

Hassan, K. (2020). 'CPEC: A Win-Win for China and Pakistan'', *Human Affairs*, 30 (2), pp. 212–223. doi: 10.1515/humaff-2020-0020.

Huaxia, (2022) 'CPEC alleviates Pakistan's energy shortage, promote renewable energy goals: official', *News.CN,* 9 September. Available at: https://english.news.cn/20220909/6ee169366e2a47b0bdb7bab3de459e09/c.html (Accessed: 22 December 2022).

Hussain, Abid. (2023) 'China to "deepen and expand" military ties with Pakistan', *Aljazeera*, 27 April. Available at: https://www.aljazeera.com/news/2023/4/27/china-to-deepen-and-expand-military-ties-with-pakistan (Accessed: 29 April 2022).

Iqbal, K. (2018). 'Securing CPEC: Challenges, Responses and Outcomes', in Arduino, A. & Gong, X. (eds.) *Securing the Belt and Road Initiative: Risk Assessment, Private Security and Special Insurances along the New Wave of Chinese Outbound Investments*. Singapore: Palgrave Macmillan, pp. 197–214.

Jadoon, A. (2018). *Allied & Lethal: Islamic State Khorasan's Network and Organizational Capacity in Afghanistan and Pakistan*. New York, NY: Combating Terrorism Center at West Point.

Khattak, D. (2018) 'The Hard Limits of Pakistan's Media Freedom', *The Diplomat*, 28 March. Available at https://thediplomat.com/2018/03/the-hard-limits-of-pakistans-media-freedom/ (Accessed: 08 October 2022).

Kilcullen, D. (2006) *'Three Pillars of Counterinsurgency'*. Available at: http://www.au.af.mil/au/awc/awcgate/uscoin/3pillars_of_counterinsurgency.pdf; https://tamilnation.org/armed_conflict/060928kilcullen.htm (Accessed: 22 January 2022).

Kugelman, M. (2023) 'Have China and Pakistan Hit a Roadblock?', *Foreign Policy*, 9 February. Available at: https://foreignpolicy.com/2023/02/09/china-pakistan-cpec-infrastructure-economy/ (Accessed: 11 March 2023).

Kupecz, M. (2012). 'Pakistan's Baloch Insurgency: History, Conflict Drivers, and Regional Implications', *International Affairs Review* (Washington, DC), 20 (3) (Spring), pp. 95–110.

Lindley, D. (2022) 'Assessing China's Motives: How the Belt and Road Initiative Threatens US Interests', *Journal of Indo-Pacific Affairs, Air University Press*, 1 August. Available at: https://www.airuniversity.af.edu/JIPA/Display/Article/3111114/assessing-chinas-motives-how-the-belt-and-road-initiative-threatens-us-interests/ (Accessed: 26 November 2022).

Mahsood, A. & Miankhel, A. K. (2013). 'Baluchistan Insurgency: Dynamics and Implications', *Global Advanced Research Journal of Social Science (GARJSS)*, 2 (3) (March), pp. 51–57.

Mangi, F. & Haider, K. (2023) 'Pakistan Launches $2.7 Billion China-Designed Nuclear Plant', *Bloomberg,* 2 February. Available at: *https://www.bloomberg.com/news/articles/2023-02-02/pakistan-launches-2-7-billion-china-designed-nuclear-plant#xj4y7vzkg?leadSource=uverify%20wall* (Accessed: 26 February 2023).

Mostofa, A. (2017) 'Politics and the press during 1971', *Dhaka Tribune,* 23 August. Available at https://www.dhakatribune.com/magazine/weekend-tribune/17336/politics-and-the-press-during-1971 (Accessed: 22 February 2022).

Nadir, M. (2022) 'Balochistan in media – Media in Balochistan', *Daily Times*, 13 June. Available at: https://dailytimes.com.pk/950969/balochistan-in-media-media-in-balochistan/ (Accessed: 13 September 2022).

Nasar, A. S. (2016) 'Balochistan: Media blackout', *Dawn,* 11 September. Available at https://www.dawn.com/news/1283382 (Accessed: 25 March 2022).

Nawaz, S. (2011). *Learning by Doing – The Pakistan Army's Experience with Counterinsurgency*. Washington, DC: The Atlantic Council.

Noorani, A. G. (2015) 'Official secrets', *Dawn,* 9 May. Available at: https://www.dawn.com/news/1180818 (Accessed: 13 September 2022).

Pakistan Today (2023) *Bilawal visit to Russia: Two countries agree to boost cooperation in diverse sectors*. Available at: https://www.pakistantoday.com.pk/2023/02/01/bilawal-visit-to-russia-two-countries-agree-to-boost-cooperation-in-diverse-sectors/ (Accessed: 17 April 2023).

Pakistani Army. (2013). *Sub Conventional Warfare (SCW) Doctrine (Publication No. AP 2601E)*. Rawalpindi: Doctrine & Evaluation Directorate, GHQ. Available at: https://www.scribd.com/document/474069737/Sub-Conventional-Warfare-Doctrine-pdf (10 August 2022)

Paul, C., Clarke, C. P., Grill, B. & Dunigam, M. (2013a). *Paths to Victory, Lessons from Modern Insurgencies*. Santa Monica, CA: National Defence Research Institute, RAND Corporation.

Paul, C., Clarke, C. P., Grill, B. & Dunigam, M. (2013b). *Paths to Victory, Detailed Insurgency Case Studies*. Santa Monica, CA: National Defence Research Institute, RAND Corporation.

Pillalamarri, A. (2015) 'The China-Pakistan Economic Corridor Is Easier Said Than Done', *The Diplomat,* 24 April. Available at: https://thediplomat.com/2015/04/the-china-pakistan-economic-corridor-is-easier-said-than-done/ (Accessed: 15 March 2022).

Priyashree, A. (2006) *Baluchistan: A Backgrounder*. New Delhi: Institute of Peace and Conflict Studies. Available at: https://www.files.ethz.ch/isn/95444/IPCS-Special-Report-32.pdf (Accessed: 15 August 2022).

Ramzy, A. & Buckley, C. (2019) 'Absolutely No Mercy: Leaked Files Expose How China Organized Mass Detentions of Muslims', *The New York Times*, 16 November. Available at: https://www.nytimes.com/interactive/2019/11/16/world/asia/china-xinjiang-documents.html (Accessed: 27 November 2022).

Rana, S. (2015) 'China-Pakistan economic corridor: Lines of development – Not lines of divide', *The Express Tribune*, 17 May. Available at: http://tribune.com.pk/story/887949/china-pakistan-economic-corridor-linesof-development-not-lines-of-divide/ (Accessed: 25 September 2022).

Sabharwal, S. (2022). *India's Pakistan Conundrum: Managing a Complex Relationship*. Abingdon, Oxon: Routledge.

Samad, Y. (2014). 'Understanding the Insurgency in Balochistan', *Commonwealth & Comparative Politics*, 52 (2), pp. 293–320. doi: 10.1080/14662043.2014.894280.

Shakil, F. M. (2022) 'Insurgents step up attacks on Chinese in Pakistan', *Asia Times*, 20 May. Available at: https://asiatimes.com/2022/05/insurgents-step-up-attacks-on-chinese-in-pakistan/ (Accessed: 10 December 2022).

Tharoor, I. (2015) 'What China's and Pakistan's special friendship means', *The Washington Post*, 21 April. Available at: https://www.washingtonpost.com/news/worldviews/wp/2015/04/21/what-china-and-pakistans-special-friendship-means/?noredirect=on&utm_term=.44f3b6deac91 (Accessed: 25 October 2022).

The Express Tribune (2013) *New doctrine: Army identifies 'homegrown militancy' as biggest threat*. Available at https://tribune.com.pk/story/488362/new-doctrine-army-identifies-homegrown-militancy-as-biggest-threat/ (Accessed: 25 March 2022).

Tiezzi, S. (2015) 'Can China's Investments Bring Peace to Pakistan?' *The Diplomat*. 21 April. Available at: https://thediplomat.com/2015/04/can-chinas-investments-bring-peace-to-pakistan/ (Accessed: 22 December 2022).

Turak, N. (2023) 'Blackouts, currency dives and corruption: Pakistan's economy is on the brink of collapse', *CNBC,* 3 February. Available at: https://www.cnbc.com/2023/02/03/blackouts-currency-dives-and-corruption-pakistans-economy-is-on-the-brink-of-collapse-.html (Accessed: 11 April 2023).

UN Human Rights Office (OHCHR), (2022). *OHCHR Assessment of human rights concerns in the Xinjiang Uyghur Autonomous Region, People's Republic of China.* Available at: https://www.ohchr.org/sites/default/files/documents/countries/2022-08-31/22-08-31-final-assesment.pdf (Accessed: 21 December 2022).

US Department of State. (2022). *2022 Investment Climate Statements: Pakistan.* Available at: https://www.state.gov/reports/2022-investment-climate-statements/pakistan/ (Accessed: 11 March 2023).

Wang, H. & Miao, L. (2016). '"Going Global Strategy" and Global Talent', in *China Goes Global.* Palgrave Macmillan Asian Business Series. Palgrave Macmillan, London. doi: 10.1007/978-1-137-57813-6_6.

Webber, L. & Valle, R. (2022) 'Islamic State Khorasan's Expanded Vision in South and Central Asia', *The Diplomat,* 26 August. Available at: https://thediplomat.com/2022/08/islamic-state-khorasans-expanded-vision-in-south-and-central-asia/ (Accessed: 20 April 2023).

Weihong, X. (2015) 'China-Pakistan Economic Corridor key to economic upgrading in southern Xinjiang', *Global Times,* 05 July. Available at: http://www.globaltimes.cn/content/930491.shtml (Accessed: 25 June 2022).

Wilkens, A. (2015) 'The Crowded-Out Conflict: Pakistan's Balochistan in its fifth round of insurgency', *Afghanistan Analysts Network,* 16 November. Available at https://www.afghanistan-analysts.org/the-crowded-out-conflict-pakistans-balochistan-in-its-fifth-round-of-insurgency/ (Accessed: 15 March 2022).

Wolf, S. O. (2017). 'China-Pakistan Economic Corridor and Its Impact on Regionalisation in South Asia', in Bandyopadhyay, S., Torre, A., Casaca, P. & Dentinho, T. (ed.) *Regional Cooperation in South Asia: Socio-Economic, Spatial, Ecological and Institutional Aspects.* Cham, Switzerland: Springer International Publishing AG, pp. 99–112.

Xinhua Silk Road (2021) *Explain China's Silk Road Economic Belt.* Available at: https://en.imsilkroad.com/p/312941.html (Accessed: 11 June 2022).

Ze, S. (2014) '"One Road & One Belt" & New Thinking with Regard to Concepts and Practice'. [*Presentation at the 30th anniversary conference of the Schiller Institute,* Frankfurt, Germany: The International Schiller Institute]. Available at: http://newpara-digm.schillerinstitute.com/media/one-road-and-one-belt-and-new-thinking-with-regard-to-concepts-and-practice/ (Accessed: 15 August 2022).

Zeb, R. (2019). *Ethno-Political Conflict in Pakistan: The Baloch Movement.* London: Routledge. https://doi.org/10.4324/9780429318139

2 Conceptual Framework

Conflict Ecosystem and Three Pillars of Counterinsurgency

After introducing the issue of the Security of the China Pakistan Economic Corridor (CPEC) and the Counterinsurgency in Balochistan in the earlier chapter, this chapter delves into the conceptual framework and research methodology deployed in the book.

Conceptual Framework

In this book, Kilcullen's (2006) three pillars of counterinsurgency model (3PCM) is the conceptual framework to help unpack the counterinsurgency strategy in light of the COIN doctrine (2013) of the Pakistani Army and thus critically analysing its efficacy in eradicating the Balochistan insurgency for establishing the safety and security of the CPEC. The main reason for using 3PCM is that it provides a relevant framework for the evaluation of a "whole-of-government" counterinsurgency strategy keeping in view the background and situation of Balochistan.

The *Three Pillars of Counterinsurgency* were conceived and delivered by Kilcullen just prior to the publishing of the US Army Counterinsurgency Field Manual (FM 3–24) in late December 2006 (US Department of Army, 2006), the panel of which Kilcullen was also a member. This was the precise time the US security establishment was busy taking initiatives for the development of a comprehensive national counterinsurgency framework and implementation plan based on an interagency approach so as to enable the United States to overthrow the threats to peace and security posed by the transnational insurgent groups in the twenty-first century. The prevailing political/security environment in Iraq and Afghanistan was such that the US government had to marshal not only its indigenous agencies (there were more than 17 agencies in the intelligence domain alone) but also the diverse variety of the host nations agencies, coalition partners in Global War on Terrorism (GWOT), international institutions, NGOs, and media, etc. Some of them had counterinsurgency doctrines, but others did not as they rejected the very idea of counterinsurgency. The Pakistani Army, a coalition partner to the United States in the GWOT, was in the latter category with no counterinsurgency doctrine at all. To create meaningful and advantageous collaboration between hosts of actors, Kilcullen came up with his easily grasped 3PCM which may be termed an "operational design" borrowing the conventional war terminology (although he referred to it as a model).

DOI: 10.4324/9781003413905-2

This model has two parts: the first pertains to the elaboration of the "conflict ecosystem" that forms the environment for twenty-first-century counterinsurgency operations (Figure 2.1), and the second is a framework for the whole-of-government counterinsurgency in that environment (Figure 2.3) (Kilcullen, 2006, p. 2). The description of the "conflict ecosystem" is similar to the prevailing environment in Balochistan (Figure 2.2), which necessitated the counterinsurgency operations by the Pakistani Army. Likewise, the three pillars framework for whole-of-government counterinsurgency (the second part of the model for inter-agency collaboration) is supported by the Pakistani Army counterinsurgency doctrine to a large extent. These two claims will be elucidated in the ensuing paragraphs. Owing to these reasons, the 3PCM is best suited to be used as the analytical/conceptual framework for evaluating the efficacy of the Pakistani Army counterinsurgency strategy in Balochistan.

The Conflict Environment

Kilcullen defines insurgency as "a struggle for control over a contested political space, between a state (or group of states or occupying powers), and one or more popularly based, non-state challengers…" (2006, p. 2). He further maintains, on the basis of his insight and experience into the "small world, scale-free" aspects of insurgent social networks and the influence of pre-war oligarchy in Iraq, "the insurgencies are fought through pre-existing social networks (village, tribe, family, neighbourhood, political or religious party) and exist in a complex social, informational and physical environment" (2006, p. 2). He makes this environment metaphor by coining the term "*conflict ecosystem*" (2006, p. 2). As mentioned in the previous chapter, the insurgent organisations in Balochistan are struggling against the state of Pakistan to make Balochistan an independent entity (Lieven, 2017, p. 179) where they can control the resources. Moreover, the Baloch society is formed on the basis of tribe/caste system with strong allegiance to the tribal chiefs. The tribes occupy various geographical areas and possess a strong affinity for their neighbourhood while still living in a complex social, informational, and physical environment. By these criteria, the prevailing social/demographic environment in Balochistan can perfectly be defined as the "*conflict ecosystem*".

Following the ecological metaphor, the conflict ecosystem consists of many independent but interlinked actors. These actors keep on evolving and adapting; some act as top predators while others seek a secure niche. Some of the actors existed before the conflict started, including government agencies, tribes, economic and political institutions, and rural populations. Others have evolved during the conflict, e.g., local armed organisations, and foreign armed individuals/groups. Moreover, the conflict generates internally displaced persons (IDPs) and refugees. As a result of the conflict, the economic dynamics of the area deteriorates to a great extent, and it results in a high crime rate, unemployment which in turn creates armed groups like black marketers, smugglers, and narcotics traffickers. This situation is graphically illustrated in Figure 2.1 (on the next page).

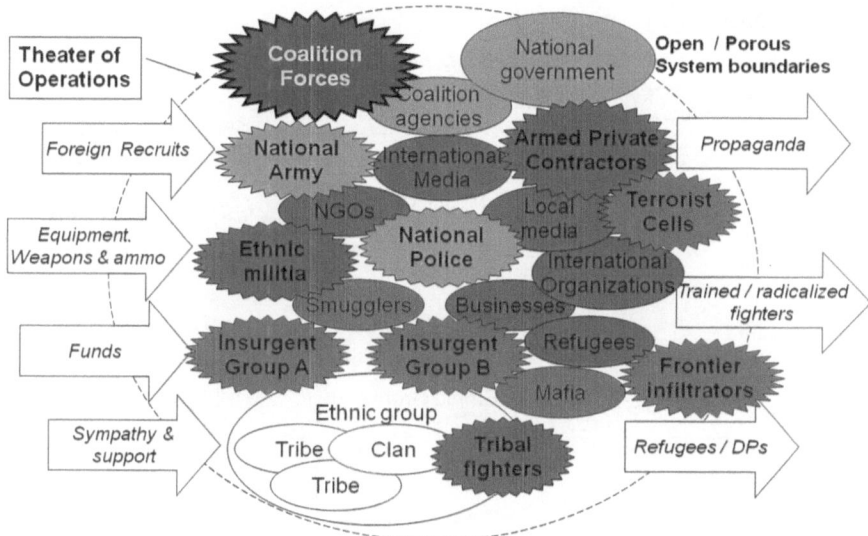

Figure 2.1 Conflict Ecosystem by Kilcullen (2006, p. 3).

In the case of the Balochistan Conflict Ecosystem, many independent but inter-linked actors exist. These include insurgent organisations such as the Baloch Libera-tion Army (BLA), Baloch Republican Army (BRA), several paramilitary forces/law enforcement agencies (LEAs) such as the Frontier Corps (FC), Police, Balochistan Levies[1], Pakistani Army, Baloch tribes, sectarian militant organisations, media and smugglers, etc. Many of these actors, such as the BLA, BRA, and some of the LEAs are constantly evolving and adapting. The Army, FC, Police, Levies, tribes, and the lo-cal government of Balochistan existed prior to the conflict in this ecosystem. The new state of the environment (insurgency) has drawn new actors, such as BLA, BRA, sec-tarian militant organisations, and more recently the Islamic State of Khorasan (ISKP), into the conflict. The insurgents are foreign-funded, and are supplied with weapons, equipment, ammunition, and hard-core guerrilla training from outside because of the porous border of Balochistan with Afghanistan and Iran (see Appendix 1 for the international border alignment). The Balochistan insurgency produced IDPs, during 2006–2013, and many people have either migrated to the province of Sindh or the southern Punjab region to avoid the conflict. The conflict ecosystem overlay graphi-cally unfolds over the Balochistan situation, as shown in Figure 2.2 (on the next page).

It is very important to understand that counterinsurgency forces like the Pakistani Army and other LEAs are inside this Balochistan Conflict Ecosystem. The theatre of operation is not an inert medium where the Pakistani Army can prac-tice its operational art. Instead in this dynamic living system, the hosts of actors change their responses to the LEAs actions and require a continuous balancing between competing requirements. The counterinsurgent force, that is the Pakistani Army, differs from the other actors on the matter of intent. The intent of the Pakistani Army is to impose a measure of control on the overall environment. In

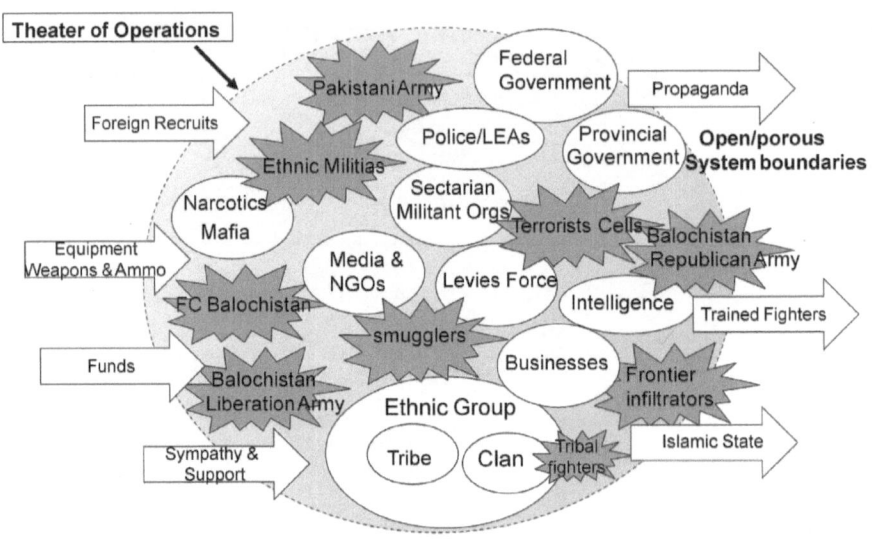

Figure 2.2 Conflict Ecosystem of Balochistan.

Source: Author.

such a complex multi-actor environment control cannot be exercised through un-questioned dominance. Instead, achieving collaboration towards a set of shared objectives with other stakeholders can help bring the situation to normalcy. In fact, marginalising and out-competing the insurgents in Balochistan to gain control over the overall socio-political space of the conflict is the main aim of the Pakistani Army (Qadeer 2022, Interview, 08 December). For this, multi-pronged efforts are required in the realm of Security, Politics, and Economy based more on the "unity of efforts" principle than the "unity of command" principle between various agencies, government, and non-government actors. The Pakistani counterinsurgency doctrine (2013) and recent practices support the idea of inter-agency counterin-surgency operations as a means of creating a shared diagnosis of the problem, platforms for collaboration, joint operations, and information sharing in order to achieve versatility, the ability to perform diverse tasks and the agility in the seam-less rapid transition between the tasks.

Three Pillars of Counterinsurgency Model (3PCM)

Kilcullen's (2006) 3PCM is designed as a base of information, three pillars (Security, Political, and Economic), and a roof of control (Figure 2.3 on the next page).

This approach "builds on 'classical' counterinsurgency theory, but also incorpo-rates best practices that have emerged through experience in peacekeeping, develop-ment, fragile states and complex emergencies in the past several decades" (Kilcullen, 2006, p. 4). Kilcullen (2006, p. 4) also maintains that this model provides the basis for measuring the progress of the counterinsurgents, and it is an aid to collaboration.

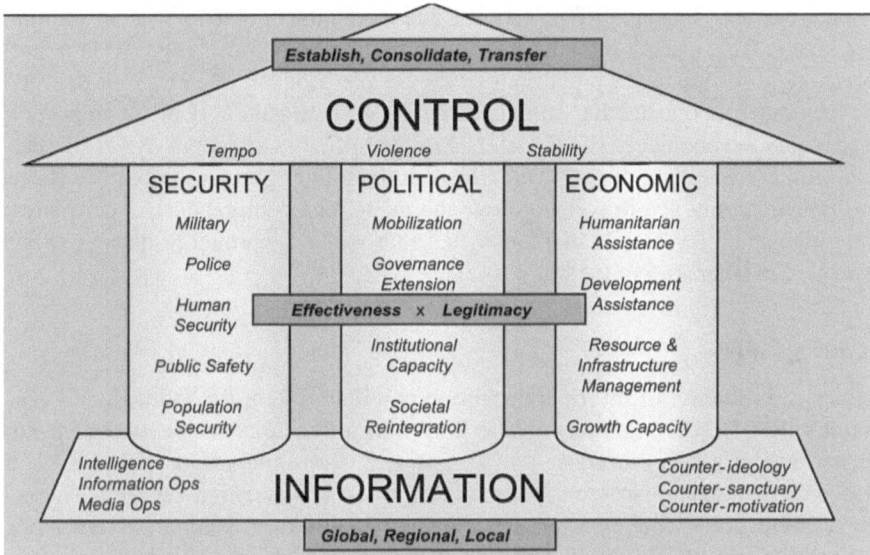

Figure 2.3 Three Pillars of Counterinsurgency: Interagency Counterinsurgency Framework by Kilcullen (2006, p. 3).

Information Base

Information is the base of all other activities within the 3PCM, owing to the fact that perception building is pivotal in establishing control and influence over the population groups (Kilcullen, 2006). Although pragmatic security, political, and economic measures are crucial in order to be effective they should be blended in the broad-ranging information strategy. Any action during a counterinsurgency campaign is meant to put across a message. The information campaign is effectively used for unifying and consolidating this message. The information campaign includes a wide array of operations such as information operations, psychological operations, electronic warfare, intelligence collection, analysis and dissemination, media operations (including public diplomacy), and countermeasures to insurgent ideology and motivation. Unless a firm base of information is developed, the three pillars of counterinsurgency cannot be effective. It is very important that the information campaign is conducted not only at the local and national level but also at the international level as modern insurgents seek the attention of the international community to gain support, funding, and sympathy. In the case of Balochistan, insurgent leaders in exile are maintaining very effective campaigns in the United Kingdom, Sweden, Switzerland, and Germany to draw upon global networks of sympathy, support, and funding (Samad, 2014, p. 308; Siddique, 2017). The Pakistani Army's counterinsurgency doctrine lays great emphasis on Information Operations (IOs) that will have a predominant role throughout the sub-conventional conflict continuum. IOs focus on all segments of society involved directly or indirectly (the neutral majority, pro or anti-state elements as well as the militants) with well-conceived narratives and counter-narratives, using different means and mediums. In other words, in addition to a kinetic fight against the adversary, both the parties are also engaged in a non-kinetic fight for the hearts and minds of the people. Both sides endeavour to erode the legitimacy of one another and influence the other to gain or deprive support of, or to harass the populace. The Pakistani Army's Military Operations (MO) Directorate, Military Intelligence (MI) Directorate and the Inter Services Public Relations (ISPR) Directorate are in collaboration and pursuing a comprehensive, consistent, and singular policy of IOs in Balochistan that will be thoroughly discussed and evaluated in Chapter 6 of the book.

Security Pillar

Resting on the base of information, the three pillars (security, political, and economic) are of equal importance. The first pillar of security in the case of Balochistan comprises the military, police, Frontier Corps (FC), and other LEAs. It also incorporates human security, public safety, and population security. The security pillar draws the attention and engages the military commander the most, but military means are applied not only in the security paradigm but all across the model. This is particularly true about Balochistan where the military spearheads all the security-related initiatives, provides assistance in economic activities/ matters (CPEC) and equally assists in political efforts, such as offering reconciliation opportunities, for marginalising insurgents. This is especially true after the commencement of CPEC projects. Therefore, military means are applied all across the 3PCM in Balochistan.

Political Pillar

Like other pillars, the political pillar also develops in the principal dimensions of effectiveness and legitimacy. The political pillar focuses on mobilising stakeholders in support of the government. The Pakistani Army's Sub-Conventional Warfare (SCW) doctrine (2013, p. 39) greatly supports the legitimacy of the overall counterinsurgency efforts by highlighting that due cognisance of legal aspects must be taken while planning and executing counterinsurgency operations. Armed Forces assist the civil authority under the provisions of Chapter 2, Article 245 of the Constitution of the Islamic Republic of Pakistan. As per the SCW doctrine (2013, pp. 23–24), the political primacy and the political aim are the fundamentals of the political prong of counterinsurgency efforts. It comprises of marginalising the insurgents and other militant groups, extending governance and promoting the rule of law by establishing the writ of the government, a key element for developing and enhancing the civil institutional capacity and effectively carrying out initiatives such as the disarming, demobilisation, and reintegration (DDR) of combatants. The Pakistani Army's doctrine supports assisting and building institutional capacity by stating that the Army, due to its inherent organisational strength, can assist the civil administration in speeding up the reconstruction process (in the conflict zone) to restore normalcy by providing specific assistance besides creating the enabling environment for initiating the process of rehabilitation. Moreover, the Pakistani Army is consistently carrying out DDR initiatives in Balochistan (Yousafzai, 2017) as a matter of practice.

Economic Pillar

The economic component of the 3PCM comprises long-term programmes for development assistance across a range of infrastructure development, industrial, and commercial activities. As Kilcullen highlights: "Assistance in effective resource and infrastructure management, including construction of key infrastructure systems, is critically important" (2006, p. 6). In the case of Balochistan, the Chinese 62 billion USD investment project of CPEC is a great economic initiative comprising diverse economic projects such as extensive rail and road infrastructure development, pipelines for oil and gas, enhancing energy (both renewable and non-renewable) capacities such as wind, hydropower energy (dams), solar, and coal (Wolf et al., 2017, p. 102). The Pakistani Army is a major stakeholder in the CPEC project because it is providing security to the project, especially in the insurgency ridden Balochistan which is a direct threat to the implementation of CPEC.

Kilcullen (2006) argues that all three pillars are of equal importance and should be developed concurrently. Without this, the counterinsurgent campaign will become unbalanced. We can gauge the progress of each pillar by measuring the efficacy/effectiveness (capability and capacity) and legitimacy. This book critically analyses the Pakistani Army's counterinsurgency campaign in Balochistan in the backdrop of the CPEC and help unpack its efficacy in the realm of sub-conventional warfare. Three major dimensions, that is Security, Political, and Economic, emerge as promoted by the Pakistani Army's doctrine as well. These elements constitute the 3PCM, comprising key causal and impacting factors contributing to the outcome of the counterinsurgency campaign and were settled upon after extensive field research work and a review of primary and secondary work pertaining to counterinsurgency (including the doctrine). The factors inherent in the model provide a comprehensive analysis of the doctrinal elements that underpin counterinsurgency and a practical assessment of the application of counterinsurgency strategy by the Pakistani Army in Balochistan. Moreover, these factors are essentially timeless and universal in insurgencies, therefore circumventing any temporal or regional restrictions on the analysis. Interestingly the 3PCM is best suited to be used as an analytical framework because the general parameters set by the model for measuring the progress of the counterinsurgents are the same points which are emphasised in the SCW doctrine (2013) of the Pakistani Army, as elaborated earlier. The SCW doctrine (2013) of the Pakistani Army will be discussed in detail in Chapter 4. In a time when insurgency has come to define the future of CPEC, this framework sheds light on this often-perplexing form of warfare and hence enable the author to contribute to the existing knowledge of the counterinsurgency policymaking in the local and global context especially for China and Pakistan in achieving the strategic goal of BRI/CPEC.

Moreover, the aim of this research is not to "test" or "refute" the *three pillars of counterinsurgency* model but instead it is applied as a conceptual framework to determine the efficacy of the Pakistani Army's strategy for eradicating the Balochistan insurgency for establishing the safety of CPEC. To put it simply, this model is used to understand the effect of counterinsurgency strategies by the Pakistani Army in Balochistan against the backdrop of CPEC and not used for theory building or hypotheses testing.

"Good and Bad COIN Practices" – Effectiveness of Various Concepts for COIN

While theorising the 3PCM, Kilcullen emphasises the importance of measuring the effectiveness and legitimacy of each pillar by stating: "In developing each pillar, we measure progress by gauging effectiveness (capability and capacity) and legitimacy" (2006, p. 5). The 3PCM does not provide any standard yardstick for gauging the effectiveness of each pillar. As this research mainly focuses on the security pillar of 3PCM, it requires a broad measure for evaluating the effectiveness of various COIN concepts adopted and manifested by the Pakistani Army on the ground in Balochistan in the light of the themes enshrined in the Sub-Conventional Warfare (SCW) Doctrine 2013.

Preceding in view, the COIN scorecard (attached as Appendix 6) developed by Paul et al. (2013, p. 249) in their study *Paths to Victory: Lessons from Modern Insurgencies* provides a suitable yardstick for characterising and evaluating the counterinsurgency practices of the Pakistani Army in Balochistan (after 2013) into "Good" and "Bad". Finally, it helps in measuring and predicting the effectiveness of the COIN practices (in case more number of good practices are adopted) or otherwise (if bad practices are greater than the good practices).

The principal line of inquiry of the study mentioned above is *"When a country is threatened by an insurgency, which counterinsurgency (COIN) approaches give the government the best chance of prevailing?"* (Paul et al., 2013, p. 1). Paul et al.'s study (2013) tested the performance of 24 core concepts for COIN ranging from classical concepts like "pacification" and "resettlement" to contemporary perspectives such as "boots on the ground". He initially selected 71 most recently resolved insurgencies for the study. Later, the figure was reduced to 59 cases of insurgencies (completed between WWII and 2010) spanning 61 countries across the globe after excluding those insurgencies where outcomes were predominantly driven by exogenous factors (like the end of apartheid and colonialism) and not by the effectiveness of the COIN strategy/forces. The analysis was carried out based on both quantitative and qualitative data. The qualitative narratives helped in the formulation of hypotheses, which were tested comparatively across cases on the basis of quantitative data. Every single case was supported by a qualitative narrative and quantitative data on almost 300 individual factors. At the analytical level of the study, the good and bad COIN practices and factors highlighted in the COIN scorecard hold true for all the core cases, without any exception, examined during the study. The COIN scorecard (Paul et al., 2013. p. 249) displays 15 good and 11 bad COIN practices and differentiates the 59 core cases into COIN wins and losses.

The general finding of Paul's study (2013) is that during the decisive phase if the score is positive by subtracting the total number of bad practices from the good ones (that is more good practices than bad) then the case was a COIN win but a COIN loss if the score is negative (that is more bad practices than good ones). Without any exception it holds true. This helps in addressing the overarching research question of the study (Paul et al., 2013, p. 2) "Which COIN approaches are more effective?"

The study finds explicitly that all the winning COIN forces score at least +2 (good-minus-bad score) and the losing COIN forces score -1 or lower (good-minus-bad score). This stands valid across all the 59 core cases. That's how the study completely discriminates these historical cases into wins and losses. Another important finding of the research is that all successful COIN campaigns realised three specific factors: "The disruption of tangible support to the insurgents, the demonstration of commitment and motivation on the part of both the government and COIN forces, and flexibility and adaptability on the part of COIN forces" (Paul & Clarke, 2016, p. ix). Thus, the study concludes, "the lesson for current or future counterinsurgents is clear: Strategies that maximise the presence of good factors and minimise the presence of bad factors are endorsed by history" (Paul & Clarke, 2016, p. 2).

Need for Superimposing the "Good and Bad COIN Practices Scorecard/Table" Over the Security Pillar of 3PCM

As mentioned earlier, the 3PCM does not provide any standard measure for even broadly gauging the effectiveness of each pillar, so the *Good and Bad COIN practices scorecard* is used for the security pillar. The *Good* and *Bad* COIN practices table will be applied purely based upon the qualitative narrative details, gathered during the field research work in Pakistan, of the fifth round of insurgency. Finally, to highlight the effectiveness of the COIN practices (security pillar) of the Pakistani Army in Balochistan, the sum of good-minus-bad will be considered. The positive value depicts the effectiveness, and the negative value will show the non-effectiveness.

The *Good* and *Bad* COIN practices table is also relevant to be applied on the fifth round of insurgency in Balochistan because Paul's set of 59 core cases includes two Pakistani counterinsurgency campaigns. The first is the counterinsurgency campaign of the Pakistani Army in erstwhile East Pakistan (Bangladesh) during 1971, and the second is the fourth round of counterinsurgency in Balochistan (1973–1978). The former was a COIN loss, and the latter was a COIN win as classified by the study (Paul et al., 2013).

Moreover, the fifth round of counterinsurgency in Balochistan meets the following criteria set for all other 59 core cases selection of the study:

a) they involved fighting between states and nonstates seeking to take control of a government or region or that used violence to change government policies b) The conflict killed at least 1,000 people over its course, with a yearly average of at least 100 c) At least 100 people were killed on both sides (including civilians attacked by rebels) d) they were not coups, countercoups, or insurrections.

(Paul et al., 2013, p. 14)

Lastly, the shift in the COIN approach of the Pakistani Army in Balochistan after promulgating the SCW doctrine in 2013 marks the initiation of a new phase as per the criteria set by the study for the phase demarcation. "A new phase was declared when the case analyst recognised a significant shift in the COIN approach, in the approach of the insurgents, or in the overall conditions of the case" (Paul et al., 2013, p. 16).

In other words, the principal value of the *Good and Bad COIN practices scorecard* once applied to this study of the Pakistani COIN in Balochistan is for the diagnostic purpose only. By no means, the scorecard guarantees its ability to predict the outcome of the ongoing fifth round of Balochistan insurgency, but it only offers a good indication as to whether the counterinsurgency efforts of the Pakistani Army are on a path likely to lead to a favourable outcome. The particular factors of the *Good and Bad COIN practices scorecard* which are present or absent in the current COIN strategy of the Pakistani Army will help this study to highlight the areas ripe for the renewed emphasis by the Pakistani Army.

Methodology

This book employs numerous methodological tools for harnessing empirical information and acquiring relevant modus operandi for presenting the research findings. Essentially the critical evaluation of arguments from interviews, coupled with the reflexive use of primary and secondary sources of data, are evaluated within a case study-based approach.

While determining the research methodology, the author considered three basic conditions as prerequisites for deciding the most appropriate social science research method as emphasized by Yin (2014, p. 9; 2018, p. 9) "(a) the form of research question posed, (b) the control a researcher has over actual behavioural events, and (c) the degree of focus on contemporary as opposed to entirely historical events". The research questions posed are more explanatory in nature and asked about the contemporary events over which the researcher has no control, so the most appropriate and advantageous research design is the "case study".

Moreover, the author has no capacity to intervene in the phenomenon that is investigated in this book. Thirdly, the given research is concurrent and not retrospective in nature, so the case study is preferred as "the relevant behaviours still cannot be manipulated and when the desire is to study some contemporary event or set of events … meaning a fluid rendition of the recent past, not just the present" (Yin, 2018, p. 12). Therefore, in this book, the "case study" design is followed to carry out the research.

There is no other insurgency comparable to Balochistan on the route of BRI where China has planned its massive investment of 62 billion USD and the infrastructure development is in progress at a high pace. Moreover, nowhere on the proposed routes of the BRI is an insurgency in its fifth round. These two characteristics of the current insurgency in Balochistan makes it distinctive and non-comparable to any other region. Therefore, the research depends on a single-case study research strategy to draw valid causal inferences.

Yet another rationale for the single case study design is the longitudinal case: "Studying the same single case at two or more different points in time" (Yin, 2018, p. 51). Yin further elaborates: "There may be pre specified time intervals, such as prior to and then after some critical event, following a before-and-after logic" (2018, p. 51). In this research, there are two time intervals of the ongoing fifth round of Balochistan insurgency – before the doctrine (2006–2012) and after the formulation of the doctrine (2013–2022) – bisected by the year 2013 when the Pakistani Army first formulated its COIN doctrine and a memorandum of understanding (MoU) was officially signed between China and Pakistan. The efficacy of the COIN strategy of the Pakistani Army in Balochistan will be evaluated before and after the formal enactment of the COIN doctrine.

In the absence of any similar cases of counterinsurgency, the study relies on the in-depth empirical investigation of the Pakistani Army's counterinsurgency strategy using an embedded single case study design, where the subunits of analyses have been incorporated. The subunits can lead to significant opportunities for extensive analyses thereby enhancing the insights into a single case study (Yin, 2018, p. 54).

As an explanation, the case study is a "concrete entity (e.g., … organisation, community, process, policy, practice, [place], or institution, or an occurrence such as a decision [or phenomenon])" (Yin, 2014, p. 237) that is the basis of the analysis. "The case in a case study is a unit of analysis [whereas] a unit lesser than the main unit of analysis, from which case study data are collected is called the embedded unit of analysis" (Yin, 2014, pp. 238/241). Variables, as the name implies, are those concepts whose value change over time, whereas observations are the values of the variables for each unit. In direct relevance to this book, the three *pillars of counterinsurgency* model harness the above-mentioned components to provide a firm ground for comprehensive analysis. The case selected is the fifth round of Balochistan insurgency. The units of analysis are the three distinct elements of Security, Political, and Economic. The variables encompassed by these units include counterinsurgents' military, political, and economic approaches and tempo; insurgent organisations' violence, strategy and tactics; and the economic/social development and assistance initiatives.

Sources of Evidence

There can be numerous sources of evidence in a case study ranging from documentation (formal studies, progress reports, memoranda, inter-office notes, operational orders, intelligence summaries, [general staff publications (GSP)], and field manuals, etc.), archival records, interviews as well as participant observations and ethnography (Yin, 2018, pp. 113–115). This books' data collection mainly employs qualitative methods to gather the empirical data for constructing an in-depth study.

Primary documents, secondary sources, and semi-structured interviews are consulted and drawn upon. Using multiple sources of evidence is vital for carrying out an in-depth study of the COIN strategy of the Pakistani Army in Balochistan in the backdrop of CPEC in its real-world context. All these sources are used for corroborating and augmenting evidence.

Documents and Archives

The primary documents used for the study are the documents pertaining to the counterinsurgency doctrine and operations of the Pakistani Army. This includes the Pakistani Army SCW Doctrine – 2013 commonly known as "the counterinsurgency doctrine" in the Pakistani Army, notes taken from the Post-War Committee's Report on 1971 War in East Pakistan and Post-Conflict Evaluation Committee's Report on Balochistan-1979, a few of the training notes and extracts from the war diaries of the Pakistani Army[2].

The primary source documents are very important in carrying out research because they offer a first-hand insight into the decision-making process and hence help in understanding accurately the representation of events. Nonetheless, the authenticity and credibility of the document(s) ought to be questioned by a researcher (MacDonald & Tipton, 1993). In other words, just acceptance of the document(s) may not lead to sound and reliable analysis unless they are authentic. Keeping this in mind, the documents mentioned above have been retrieved through open web sources. These documents have been cross-checked for authenticity and originality from the different sets of serving and retired officers of the Pakistani Army. These documents are essential for gaining a strong foundation of primary source knowledge on the efficacy of the counterinsurgency strategy (2013–2022) of the Pakistani Army in Balochistan.

The second source of evidence is an array of secondary sources like books, newspaper articles, research journal articles, think tank reports, media news (including social media) which are produced as the result of an extensive literature search. The secondary sources will help in augmenting the study by explaining the theoretical, doctrinal, conceptual, and security policy related concepts. Moreover, the secondary resources will be of great help by providing data that has not been acquired by the author. These data will ultimately be used for the triangulation of the data gleaned from multiple other sources by the author. The quantitative data is also gathered from various open access sources. The details of the secondary sources for the quantitative data are given in Table 2.1.

Table 2.1 Sources of Data

Sources	Status of Data
Home Department Government of Balochistan/ Office of the Chief Capital Police Officer (CCPO), Quetta, Balochistan	Data obtained via personal liaison by the author
South Asia Terrorism Portal (SATP)	Open-Source Data
Uppsala Conflict Data Program (UCDP)	Open-Source data

Each source highlights diverse data such as the incidents of terrorism in Balochistan claimed by some insurgent organisation, the number of surrendered insurgents, and the number of fatalities in the armed clashes between the LEAs and insurgents, etc. The brief information about the sources of the data is as follows.

Home Department Government of Balochistan/Office of the Chief Capital Police Officer (CCPO), Quetta, Balochistan

Data taken from First Hand Information Reports (FIRs) launched in police stations across the province was obtained about the insurgents/terrorist incidents in the province from 2006–2022 claiming human lives and any loss to infrastructure. Other relevant details, such as the claiming of responsibility for the incident and the arrest of any insurgents were also obtained. Moreover, data was collected on the numbers of insurgents who opted for the disarming, demobilisation, and reintegration (DDR) initiative of the government. The central police record office data was also accessed.

South Asia Terrorism Portal (SATP)

SATP is the largest website on terrorism and low-intensity warfare in South Asia and creates the database and analytic context for research and analysis of all extremist movements in the region.

Uppsala Conflict Data Program (UCDP)

The Uppsala Conflict Data Program (UCDP) is a comprehensive and publicly available project of collecting information on armed conflicts by Uppsala University in Sweden. UCDP includes both qualitative and quantitative data.

Semi-structured Interviews

One of the most important sources of evidence in the case study is interviews (Yin, 2018, p. 118) because most case studies are about human affairs or actions. Specifically, in this book, interviews are the mainstay as there is little published literature on the Balochistan security situation or its connection to the CPEC. There is also a large void on the current round of insurgency in Balochistan and corresponding COIN strategy of the Pakistani Army as amply highlighted in the Introduction chapter. In this study, the semi-structured interview technique was adopted as it provides a more flexible interview process. By following Morse (2012, p. 197), the "interviews consisted of the question stem to which participant could respond freely; probing questions planned and arising from the participant's response were asked". This technique is also capable of unveiling significant and often concealed aspects of human and organisational behaviour, and thus it is the most effective means of collecting information (Kvale & Brinkmann, 2009). In this study, semi-structured interviews help in disclosing important and often hidden facets of the

Pakistani Army's organisational behaviour such as how it identified the need for developing counterinsurgency doctrine and how the organisation's conventional wisdom developed and in turn helped adopt successful or non-successful COIN practices in Balochistan. Moreover, the interviews give a good insight into first, the various dynamics of insurgency from the perspective of the Baloch people who favour the cause of an independent Balochistan and second, the existing law and order situation in the province.

The interviews were conducted in two phases as part of field research work in Pakistan. The first phase was between December 2018 and February 2019, and the second phase resumed in 2022. During the period between the two rounds of interviews, there were frequent lockdowns caused by the global pandemic (Covid-19), and afterwards, the author was unable to visit Balochistan due to security concerns. The interviews included various individuals directly linked to CPEC and its security, such as serving and retired army officers, Provincial Government Officials of Balochistan, prominent Baloch journalists, rights activists in Balochistan, and the leaders of the Balochistan nationalist political parties. All the interviews were conducted one-on-one. A total of 30 interviews were conducted, dividing the interviewees into three distinct groups: Military personnel, civilian executives/government officials and the Baloch sub-nationalists/local population. Three Master Question Lists were prepared, one for each group, as this helped in pursuing a consistent line of inquiry as reflected by this case study's protocol but at the same time, the questions in the interviews were fluid rather than rigid. Verbalising the questions in an unbiased manner was another essential aspect that was taken care of during the interviews. The specific questions were carefully worded in an unbiased manner so as to avoid leading questions which may have influenced the response of the interviewees.

Although the aim of the interviews was to acquire the opinions based on how respondents construct reality, it is imperative that the interviewees had considerable practical experience of counterinsurgency in Balochistan or exposure to the prevailing environment in the province (in case of civilians) by virtue of serving/living there for a long time to validate their opinions on the subject. The Purposive Sampling Technique[3] (Lewis-Beck, Bryman & Futing Liao, 2004; Marshall, 1996; Ritchie et al., 2013) was used. The military/LEAs personnel were selected based on their direct involvement in tackling the insurgency either in the planning process or actively participating in the combat operations in Balochistan. Some of the military officers were chosen because of their direct role in the formulation of the SCW doctrine (2013) at Doctrine and Evaluation (D&E) Directorate of the Pakistani Army and the implementation of the new training regimes in the light of the SCW doctrine while serving at Inspector General Training and Evaluation (IGT&E) Branch. The civilian government officials and executives were selected if they fulfilled the following two conditions; firstly, having domicile of origin in Balochistan and secondly serving for a minimum for ten years in the province[4]. Lastly, the Baloch sub-nationalist leaders and other people from different spheres of the civil society of the province, who were not government employees, were selected if they fulfilled the criteria of living in the province for a minimum of ten years[5]. Moreover, a Snowball Sampling Technique[6] (Lewis-Beck et al., 2004; Parker, Scott, & Geddes, 2020) was also employed, and many of the interviewees were selected based on the recommendation of the first participants, provided that they fulfilled all the specific criteria set for each group, sampled up or leveraging on the connections and network of former colleagues in the Pakistani Army. The broad diversity of the interviewees aligns well with the study where there was a requirement of obtaining rich, reliable, and robust evidence/data. The evidence from the interviews, which are direct quotations from the respondents, were also used in addition to content analysis of relevant documents to support the answers to the research questions.

Special attention has also been paid to the ethics of conducting this research. None of the interviewees consented to the digital recording of the interviews, but permission was given for note-taking. Moreover, almost all the interviewees, except Rahimuddin Khan, strictly laid down the condition of anonymity owing to the harsh censorship policies of the government in Pakistan. Therefore, the participants' names have been anonymised less Rahimuddin Khan alias *Rahim, the brave & the bold*. These semi-structured interviews have some strengths and weaknesses in terms of fulfilling expectations. The foremost strength of these interviews is that they provided a wealth of informal data and insight about the Pakistani Army's way of learning and conducting the COIN campaign and the subsequent changes in the COIN strategy after the promulgation of the doctrine. Moreover, these interviews greatly reflect the civilians of Balochistan's point of view about the CPEC, government policies, security situation, and concerns about the future of the province against the backdrop of the CPEC. Furthermore, these interviews support the data revealing the institutionalisation of learning and corresponding COIN strategy in Balochistan, which is something that cannot be found in the primary and secondary sources. These interviews have some shortcomings as well. For instance, in a few of the interviews of the Army personnel and Government Officials, it seems that participants' answers are depicting the official stance of the Government/Army, rather than representing an objective assessment. Consequently, these interviews have been triangulated with primary and secondary sources, where available, in recognition of the issues of bias, poor recall, and the subjectivity of interview data.

Thematic Analysis and Triangulation of Data from Multiple Sources of Evidence

The study uses thematic analysis as a method for identifying, analysing, and reporting patterns (themes) within the semi-structured interview dataset in order to organise and describe the dataset in (rich) detail. By following Braun and Clarke (2006), a theme is counted as something important about the data in connection to the research questions and that it also shows some degree of patterned response or meaning within the dataset. Moreover, its "keyness" is irrespective of being dependent on quantifiable measures. For example, in the study, "winning hearts and mind" strategy is a theme that led to the considerable decrease in insurgent activities in Awaran District after it was adopted by the Pakistani Army in 2016.

For each individual theme/subtheme, a detailed analysis was written as well as identifying the "story" that each theme tells in connection to the research questions. Finally, the analysis of the fully worked out themes forms the basis of the empirical chapters of this book.

Data triangulation conveys validity into the case study. As Stakes maintains: "By adopting a triangulated approach, researchers are using multiple techniques to clarify meaning" (2005, cited in Hesse-Biber, 2017, p. 230). In this study, the data triangulation technique is used for building the validity of the case study. Data, as mentioned above, has been collected from various sources for corroborating and augmenting pieces of evidence. In essence, data from multiple sources, including the interviews, documents, and secondary sources of quantitative data, was compared and contrasted to either substantiate or refute an idea.

Ethical Considerations

The book deals with the sensitive topic of insurgency, which also comprises of semi-structured interviews of various individuals directly linked to CPEC and its security as part of field research work. Therefore, the ethical considerations are of paramount importance.

The foremost ethical consideration practised was gaining informed consent from all the interviewees. Moreover, the interviewees were extended the right to place restrictions on digital/manual recording. As mentioned earlier, protecting the privacy and confidentiality of the participants was extremely important keeping in view the strict censorship policies of the government and crackdown in the country on dissenting voices by the military establishment/intelligence agencies (Kermani, 2021). The participants are referred to by their allocated pseudonyms except Rahimuddin Khan. In the book, no personal details allowing identification of the participants are provided except the one. The only information about interviewees is regarding his/her profession, affiliated organisation, and their views.

Limitations of the Research

There are a couple of issues which the author faced during the field research work. For instance, owing to the strict censorship policies of the current government in Pakistan, many people were generally reluctant to share their opinions. Those who agreed to give an interview strictly laid down the condition of anonymity.

The author was unable to maintain gender parity amongst the interviewees. All the interviewees are male except one social activist (Shazia Haider) from Balochistan, who is female. The main reason for the void of female interviewees is the tribal culture of Balochistan, where women are strictly not supposed to talk to a stranger (Zafar, 2016, p. 25, 30, 174; Shazia Haider, 2019, Interview, 11 January; Kakar & Synovitz, 2020). Even the women working in government offices like the Balochistan Civil Secretariat avoid talking to any male who is not a member of the office staff. That is why author could not interview females from the civil society of Balochistan. No female members of the armed forces were interviewed because women do not serve in the fighting units of the army, and hence they are not part of the operational planning process in any way. Therefore, the author was unable to include any woman apart from Shazia Haider on the list of interviewees.

Moreover, the author could not get the data from some of the provincial civil departments of Balochistan, such as Balochistan Technical Education and Vocational Training Authority (B TEVTA) and the Civil Secretariat about insurgent vocational training courses and exact employment figures of the former insurgents. Again, the reason was the fear in the people of violating the strict censorship policy of the government.

The statistical data of the activities of the ISKP in Balochistan was not available to be used in the study. The Capital Chief Police Officer (CCPO) Office denied having any such data. The reason is that officially the Government of Pakistan and the Pakistani Army deny the presence of the ISKP in the country. Moreover, the data about ISKP activities in Pakistan is not available on any of the open or restricted access online datasets like SATP, UCDP, etc. – although some of the devastating suicide attacks by ISKP in Balochistan, as reported in the news media, have been frequently referred to in the study.

Notes

1 A provincial paramilitary force whose main task is the maintenance of law and order in the province of Balochistan (Kundi, 1993, p. 26).
2 These documents are available on academic material websites or online digital libraries such as Scribd®, Academia.edu®, and SlideShare®, among others. All the documents regarding the Pakistani Army referred to in the book are accessed strictly through the internet (open web sources) only.
3 Purposive sampling is a non-probability sampling technique. In qualitative inquiry, purposive sampling is the deliberate seeking out of participants with particular characteristics, according to the needs of the developing analysis (Lewis-Beck et al., 2004; Marshall, 1996; Ritchie, Nicholls & Ormston, 2013).
4 In this study, one of the criteria of participants (civilian government officials of Balochistan) was their service in the province for a minimum of ten years to enhance the credibility of the interviewees. The ten years' criterion was set keeping in view the Government of Pakistan's policy about the stay of the civil gazetted officers in Balochistan. According to the policy, these officers are rotated in all the province during their service. The author considered ten years as a reasonable period for participants' familiarity with the Balochistan context. Somebody who had served in Balochistan for an extended tenure of ten years was able to comment on the situation of the province in a more informed way.
5 Living for a minimum of ten years in Balochistan was set as a criterion for the people from the civil society to enhance the credibility of the interviewees. Balochistan is a backward area where most of the people (adult males) from civil society, mostly traders and businessman, spend time in other parts of the country such as Karachi and Lahore for establishing the businesses for better earnings or doing various jobs. They segregate their time between Balochistan and the other developed areas of Pakistan. Somebody who had lived in Balochistan for ten years was in a better position to comment on the socio-political and security situation of the province because of the ample opportunities of social interactions, attending political rallies etc. in the province.
6 In qualitative research, snowball sampling is one of the most popular techniques of sampling. The characteristics of networking and referral are central to the snowball

sampling technique. Initially, the author starts with a small number of research subjects who fulfil the research criteria. These initial participants are then asked to recommend other contacts, potential research participants, who also fit the research criteria who then, in turn, recommend other potential participants, and so on (Lewis-Beck et al., 2004; Parker et al., 2020).

References

Braun, V. & Clarke, V. (2006). 'Using Thematic Analysis in Psychology', *Qualitative Research in Psychology*, 3 (2), pp. 77–101. ISSN 1478-0887.

Hesse-Biber, S. N. (2017). *The Practice of Qualitative Research*. Thousand Oaks, CA: Sage.

Huma Zafar. (2016) *Empowering Women: NGOs Project Impacts in Baluchistan-Pakistan*. PhD Thesis. Western Sydney University, Australia. Available at: https://www.google.com/url?sa=i&rct=j&q=&esrc=s&source=web&cd=&cad=rja&uact=8&ved=0CAMQw7AJahcKEwio64r6_7f-AhUAAAAAHQAAAAAQAw&url=https%3A%2F%2Fresearchdirect.westernsydney.edu.au%2Fislandora%2Fobject%2Fuws%3A37905%2Fdatastream%2FPDF%2Fdownload%2Fcitation.pdf&psig=AOvVaw0mbsxoos8HVp_JQqitjqJD&ust=1682064297996250 (Accessed: 11 September 2022).

Kakar, A. & Synovitz, R. (2020). 'Pakistani Man Is Blinded by His Father, Brothers for Wanting a Love Marriage', *Radio Free Europe Radio Liberty (RFERL)*, 22 January. Available at: https://www.rferl.org/a/pakistani-man-blinded-by-his-father-brothers-for-wanting-love-marriage-/30391592.html (Accessed: 26 July 2022).

Kermani, S. (2021). 'Pakistani intelligence accused of torture in crackdown on dissent', *BBC News,* 02 June. Available at: https://www.bbc.com/news/world-asia-57241981 (Accessed: 11 September 2022).

Kilcullen, D. (2006) *'Three Pillars of Counterinsurgency'*. Available at: http://www.au.af.mil/au/awc/awcgate/uscoin/3pillars_of_counterinsurgency.pdf; https://tamilnation.org/armed_conflict/060928kilcullen.htm (Accessed: 22 January 2022).

Kundi, M. A. (1993). *Balochistan, a Socio-Cultural and Political Analysis*. Pakistan: Qasim Printers.

Kvale, S. & Brinkmann, S. (2009). *Interviews: Learning the Craft of Qualitative Research Interviewing*. Los Angeles, CA: Sage.

Lewis-Beck, M. S., Bryman, A. & Futing Liao, T. (2004). *The Sage Encyclopaedia of Social Science Research Methods*. Thousand Oaks, CA: Sage Publications, Inc.

Lieven, A. (2017). 'Counter-Insurgency in Pakistan: The Role of Legitimacy', *Small Wars & Insurgencies*, 28 (1), pp. 166–190. doi: 10.1080/09592318.2016.1266128.

MacDonald, K. & Tipton, C. (1993). 'Using Documents', in Gilbert, N. (ed.) *Researching Social Life*. London: Sage, pp. 195–198.

Marshall, M. N. (1996). 'Sampling for Qualitative Research'. *Family Practice*, 13(6), pp. 522–526. doi: https://doi.org/10.1093/fampra/13.6.522.

Morse, J. M. (2012). 'The Implications of Interview Type and Structure in Mixed Methods Designs', in Gubrium, J. F., Holstein, J. A., Marvasti, A. B. & McKinney, K. D. (eds.) *The SAGE Handbook of Interview Research: The Complexity of the Craft*. Thousand Oaks, CA: Sage, pp. 193–204.

Pakistani Army (1979). *Post-Conflict Evaluation Committee's Report on Balochistan-1979*. Rawalpindi: Chief of General Staff (CGS) Secretariat, GHQ.

Pakistani Army. (2013). *Sub Conventional Warfare (SCW) Doctrine (Publication No. AP 2601E)*. Rawalpindi: Doctrine & Evaluation Directorate, GHQ. Available at: https://www.scribd.com/document/474069737/Sub-Conventional-Warfare-Doctrine-pdf (Accessed: 10 August 2022)

Parker, C., Scott, S. & Geddes, A. (2020). 'Snowball Sampling' in Atkinson, P.; Delamont, S.; Cernat, A.; Sakshaug, J. and Williams, R.A (ed.) *SAGE Research Methods Foundations*. Thousand Oaks, CA: SAGE Publications. Available at: http://methods.sagepub.com/foundations/snowball-sampling (Accessed: 02 August 2022).

Paul, C. & Clarke, C. P. (2016). *Counterinsurgency Scorecard Update: Afghanistan in Early 2015 Relative to Insurgencies Since World War II*. Santa Monica, CA: National Defence Research Institute, RAND Corporation.

Paul, C., Clarke, C. P., Grill, B. & Dunigam, M. (2013). *Paths to Victory, Lessons from Modern Insurgencies*. Santa Monica, CA: National Defence Research Institute, RAND Corporation.

Ritchie, J., Lewis, J., Nicholls, C. M. & Ormston, R. (2013). *Qualitative Research Practice: A Guide for Social Science Students and Researchers*. London: Sage Publications.

Samad, Y. (2014). 'Understanding the Insurgency in Balochistan', *Commonwealth & Comparative Politics*, 52 (2), pp. 293–320. doi: 10.1080/14662043.2014.894280.

Siddique, A. (2017) 'Pakistan's Balochistan Conflict Reverberates in Europe', *Gandhara*, Available at: https://gandhara.rferl.org/a/pakistan-balochistan-europe-seperatists-activists/28902514.html (Accessed: 13 December 2022).

US Department of Army. (2006). *Counterinsurgency*, Field Manual (FM) 3–24. Available at https://www.hsdl.org/?abstract&did=468442 (Accessed: 12 February 2022).

Wolf, S. O. (2017). 'China-Pakistan Economic Corridor and Its Impact on Regionalisation in South Asia', in Bandyopadhyay, S., Torre, A., Casaca, P. & Dentinho, T. (ed.) *Regional Cooperation in South Asia: Socio-economic, Spatial, Ecological and Institutional Aspects*. Cham: Springer International Publishing AG, pp. 99–112.

Yin, R. K. (2014). *Case Study Research: Design and Methods* (5th ed.). Thousand Oaks, CA: SAGE Publications, Inc.

Yin, R. K. (2018). *Case Study Research and Applications: Design and Methods* (6th ed.). Thousand Oaks, CA: SAGE Publications, Inc.

Yousafzai, G. (2017) 'Pakistan says over 300 Baloch separatist militants surrender', *Reuters (UK)*, 10 December. Available at: https://uk.reuters.com/article/uk-pakistan-militants/pakistan-says-over-300-baloch-separatist-militants-surrender-idUKKBN1E40L6 (Accessed: 10 August 2022).

3 Institutional Learning Leading to the Historical Development of the Sub-Conventional Warfare Doctrine of Pakistani Army (1947–2013)

Introduction

The Sub-Conventional Warfare (SCW) Doctrine of the Pakistani Army was promulgated in 2013. It is a key document on which the counterinsurgency strategy of the army is based (Asim 2019, Interview, 25 January; Qadeer 2022, Interview, 08 December). The Pakistani Army inherited a plethora of doctrines from the British Indian Army in 1947 after the partition of the Indian Sub-Continent which were very much in vogue till 1963 once the army published their Frontier Warfare pamphlet (Adnan 2019, Interview, 14 January). Since Pakistan's Independence in 1947, the Pakistani Army has been involved nearly continuously throughout the length and breadth of the country in fighting insurgency and rebellion. The results of the counterinsurgency campaigns of the Pakistani Army have been mixed. It quelled the rebellion in Kalat, Balochistan (1948), in FATA (1952–53 and 1954) led by Faqir of Ipi[1], but lost erstwhile East Pakistan in 1971. Between 1973–78, it crushed the insurgency in Balochistan. After 9/11, it fought a counterinsurgency campaign against the Taliban/Al Qaida in FATA/NWFP as part of the "Global War on Terror" (GWOT). Initially, the results of the campaign were not very encouraging, but then from 2008 onwards, the army learned the art of fighting insurgency the hard way under the leadership of General Kiyani (Kamran 2018, Interview, 15 December; Nadeem 2018, Interview, 13 December). Currently, the Pakistani Army is fighting the fifth round of insurgency in Balochistan (Adeel, 2015, p. 124; Sabharwal, 2022, p. 64; Wilkens, 2015).

This rich experience of fighting various counterinsurgency campaigns helped the army to learn through a systematic institutional learning process to carve the best practices and approaches in the prevailing environment. That is how it built upon its strong inherited ethos of the British Indian Army and its culture of fighting *small wars* which the Pakistani Army carried forward and frequently displayed in various counterinsurgency campaigns as late as 1973–78. However, the institutional learning cycles were time and again interrupted.

During the various learning cycles of the Pakistani Army, there were certain individual officers/field commanders such as Brigadier Akber in 1948 and Lieutenant Colonel Mitha in the early 1960s who have been instrumental and efficient

DOI: 10.4324/9781003413905-3

at kick-starting the learning process after recognising the performance gaps (Muzammil, 2018, Interview, 27 December). This corresponds to the initiation of Downie's model (first two steps) as it "begins with members of the organisation recognizing that there are performance gaps that can only be redressed through adaptation and change" (Shultz, 2012, p. 18). Their agency pushed the learning forward, but structural factors stopped it from being fully implemented. However, the predominant factors that break the institutional learning cycles have been huge structural factors. The top of the list of these structural factors is interstate war, namely the conventional warfighting focus of the Pakistani Army towards India and the major wars between the two countries. In all the first three broken learning cycles of the Pakistani Army, this was the fundamental reason for interruption in unconventional warfighting (counterinsurgency) learning. Apart from this, there were certain other structural factors which impeded the learning, such as the geostrategic priorities of the Pakistani Army during the Cold War era and later in the GWOT. The Downie's Institutional Learning Cycle and all these structural issues which hampered the learning and hence led to the non-implementation of the counterinsurgency learning will be thoroughly reflected upon later in the chapter.

This chapter traces the evolution of the Sub-Conventional Warfare (SCW) Doctrine of the Pakistani Army through the institutional learning process. Basically, the doctrine is the synthesis of the Pakistani Army's historical base of the ethos of fighting *small wars*, inherited from the rich experience of the British Indian Army, over which it developed its own diverse COIN experience while fighting in Balochistan, East Pakistan and FATA/NWFP over a period of sixty years. In other words, it is the essence of the *small wars*/counterinsurgency experience – moulded and innovated as per the modern requirements of the Pakistani Army through a systematic and analytical process.

The SCW doctrine (2013) sets the counterinsurgency approach of the modern-day Pakistani Army (Asim 2019, Interview, 25 January). This approach is traced in the following chapter, to set the firm foundations on which the later empirical chapters of the book will be based in order to evaluate two aspects: Firstly, the extent to which the Pakistani Army adheres to the doctrinal approach, as enshrined in the SCW document, by incorporating it into the COIN strategy in Balochistan; secondly, the efficacy of the COIN strategy of the Pakistani Army in fighting insurgency in Balochistan after 2013 in regard to the safety of China Pakistan Economic Corridor (CPEC). That is how this chapter and the next one, *"Institutionalisation of Learning and the Themes of the Sub-Conventional Warfare Doctrine"*, deal with the following sub research question of the study: *What is the Doctrinal Approach of the Pakistani Army towards Counterinsurgency?*

This chapter unfolds firstly by tracing the Pakistani Army's culture of fighting *small wars* which it inherited in 1947 from the British Indian Army. Then it looks at the step-by-step development of the institutional learning of the Pakistani Army towards counterinsurgency, using Downie's Institutional Learning Cycle, also highlighting the *Lessons Learned* during various counterinsurgency campaigns/periods which depict the Pakistani Army's flexibility and willingness to learn, as Nagl maintains:

> Responsive, flexible military institutions often publish "Lessons Learned" notes, incorporating information gained locally during the course of a conflict ...... Such cases should be accepted as indicators of the flexibility of the organization and its willingness to learn; however, the changes are generally incorporated into published doctrine at the first opportunity.
>
> (2005, p.7)

What Is Doctrine?

Beyond the specifics of counterinsurgency, the study of military doctrine constitutes an important part of the security studies spectrum. Doctrine is very important for understanding the causes, conduct, and aftermath of war as it is an integral element of the military component of grand strategy. Doctrine can also influence the political and economic behaviour of states. Moreover, doctrine may exert an influence on civil-military relations and alliance formation (Long, 2016, p. 7). In spite of its significance, until recently, the study of military doctrine was not homogenised into political science and international relations (Long, 2016, p. 7).

What is doctrine? It is a simple, but an important question as the answer will highlight its significance by establishing the function and role which doctrine plays for the military. In 1926, J.F.C Fuller put forward the definition of doctrine as:

> The central idea of an army ... which to be sound must be principles of war, and which to be effective must be elastic enough to admit mutation in accordance with change in circumstance. In its ultimate relationship to the human understanding this central idea or doctrine is nothing else than common sense – that is, action adapted to circumstance.
>
> (p. 254)

Since then, very little has come up to contest Fuller's point of view. Interestingly, decades after Fuller posited the definition as mentioned above, the main doctrine publications/manual of military definitions of the United Kingdom, United States, and Pakistani Army all use Fuller's concept, with little deviation from his original words, while describing doctrine as follows: "Fundamental principles by which the military forces or elements thereof guide their actions in support of national objectives. It is authoritative but requires judgment in application" (UK Army Doctrine Primer, 2011, pp. 2–1; Dictionary of Military and Associated Terms (US Department of Defence, 2016, p. 71); Pakistani Army's General Staff Publication[2] (GSP) – 1840, Glossary of Military Terms and Definitions, 1990, p. 159). The principles can vary from the policies and standard operating procedures (SOPs) in vogue in the specific branch of the military to the tactics and techniques instructed during military training to the young inductees. As Lange maintains: "A military doctrine can also be defined as the operational handbooks or tactical regulations prepared mainly by military specialists as guidance for the officer corps and troop training" (1992, p. 1). It is clear from all these definitions that the doctrine is the refinement of war experience and military history. It articulates perennial principles and systematises best practices that are supposed to be applied sensibly as per the conditions of the prevailing environment. As the UK Army Doctrine primer puts it succinctly: "Doctrine is not just what is taught, or what is published, but what is believed" (2011, p. ix).

These principles reflect the military's view about what works best in the war based on institutional learning over the years through actual wars, war gaming exercises, and successful operational planning. Principles are not checklists or the compelling set of rules but a common frame of reference to inculcate a drive in the soldiers to be creative and adaptive problem solvers. They provide the requisite flexibility for incorporating new technologies, ideas, and organisational designs (Spencer, 2016). Moreover, the doctrine also comprises tactics/techniques, SOPs, and drills (collectively abbreviated as *TSDs*) by incorporating the military's evolving knowledge and experience. TSDs help in implementing principles by linking them to application.

Military doctrine is far more than principles. It also reflects how the military intends to fight, based on its past experiences, by establishing a common frame of reference to solve military problems (Spencer, 2016). A doctrine greatly affects the posture and organisation of the military, along with its internal dynamics of training and command (Bono, 2004, p. 439). Military doctrine is informed by institutional learning and organisational culture.

Effective doctrine is not developed in a vacuum as the doctrine writers take fair account of the historical experience of the organisation, contemporary environments, modern technology, and weapons systems so that once it is promulgated it should reflect updated/relevant military knowledge. Mäder (2004, pp. 27–28) referred to it as the "collective perception of historical experience and current interests". Published doctrine displays the Army's established view though it might be surpassed owing to the swift changes in the field. The doctrine has an impact on the organisational culture, which may lead to bias in the development of doctrine because generally, the doctrine exhibits in what manner the organisation views itself as a whole. Therefore, it is very important to ascertain whether any bias existed during the time the doctrine was written (Dewar, August & Builder, 1996, p. 28; Cassidy, 2006).

For successfully carrying out military operations, the army requires an understanding based on intellectual rigour, clear articulation, and experience. Doctrine, designed to guide, not compel, provides this strong basis of understanding and of course, it needs to be applied judiciously. When the doctrine is imparted, adapted, and applied befittingly as per the changing conditions, it endures. Though doctrine is based on the experiences of war, it evolves as the result of changes in the security milieu, technological advances, and renewed experiences of contemporary operations and military wargaming. "As such, it has three functions: philosophical, based on enduring lessons from the past; practical, making it relevant to contemporary operations; and predictive, taking in foreseeable developments in terms of threats, technology and domestic policy" (UK Army Doctrine Primer, 2011, pp. 1–2). The principal purpose of the military doctrine is the provision of the framework of guidance for the successful and smooth conduct of operations in line with the prevailing environment and developments while also situating the context in which those actions will be undertaken.

Organisational Culture, Institutional Learning, and Doctrine – Downie's Cycle

Besides the historical context, a very important element which has a direct bearing on the evolution of doctrine is the organisation or institution itself – in this particular case, the Pakistani Army. As doctrine is usually conceived and designed to meet organisational requirements, so the development/evolution of doctrine is greatly influenced by the culture of the organisation. Downie (1998) pushed this forward as a central theme, while he analysed the relationship between experience, institutional learning, organisational culture, and doctrinal development. He observed a close connection between doctrinal development or evolution and the organisational culture. He identified this relationship as the main determining influence on the development of the conventional wisdom of an organisation and subsequently the way the organisation's institutional memory influences further response to change. In case the organisational culture is ambivalent to change, there is no possibility of doctrinal developments. Conversely, if the organisational culture supports change, a main amendment or revision to the doctrine is possible. On the basis of his findings, Downie describes the learning process in terms of an institutional learning cycle by defining institutional learning as:

> A process by which an organisation. .... uses new knowledge or understanding gained from the experience or study to adjust institutional norms, doctrine and procedures in ways designed to minimize previous gaps in performance and maximize future successes
>
> (1998, p. 22).

Downie maintains that adaptable institutional memory is a "prerequisite for learning and occurs when an organization captures and institutionalises lessons learned by its members" (1998, pp. 22–23). The organisation cannot learn unless it "first act(s) to interpret, evaluate, and accept the lessons learned by individual, organisational members and then make(s) the decision to adapt organizational behaviour to this new knowledge and transmit it throughout the organization" (1998, p. 23). Therefore, effective adaption is only possible if the organisation leaders are willing and have the capacity to encapsulate the lessons learned by the experiences of the organisation. Developing consensus is the cornerstone of this whole process. As Bo Hedberg puts it, "Members come and go, and leadership changes, but organizations' memories preserve certain behaviours, mental maps, norms, and values over time" (1981, cited in Nagl, 2005, p. 7). An army codifies its institutional memory in doctrine (Nagl, 2005, p. 7).

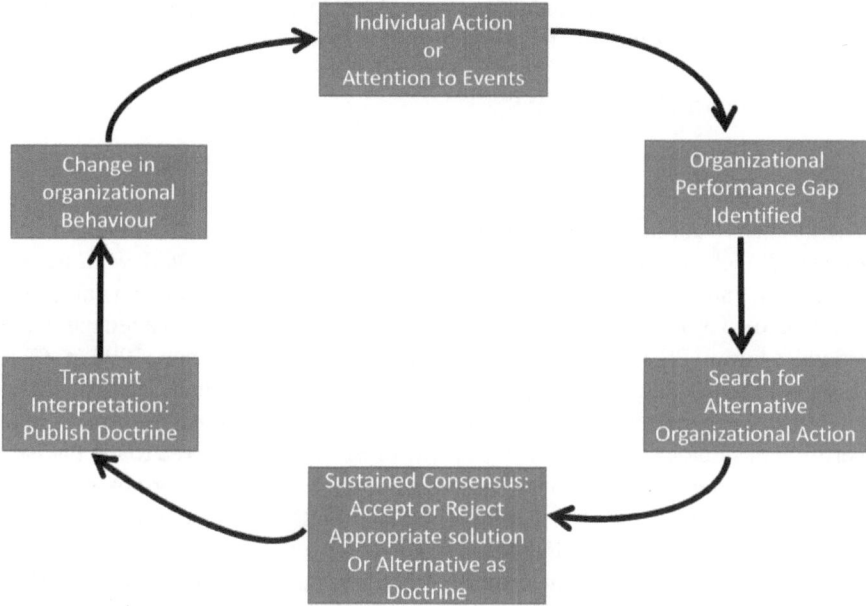

Figure 3.1 Downie's Institutional Learning Cycle: The Process of Doctrinal Development.

Downie proposed a six-step process: Starting from the individual actions, or attention to events, followed by identifying an organisational performance gap; a search for alternative organisational actions; the procedure for attaining a sustained consensus; publishing the new doctrine; and finally, the change in organisational behaviour. The cycle repeats itself (Nagl, 2005). Nagl (2005, p. 8) has further developed Downie's cycle in *Learning to Eat Soup with a Knife* which is illustrated in Figure 3.1 (above). This cycle will be used in the chapter to identify the doctrinal development stages of the Pakistani Army, which inherited the doctrinal elements of the British Indian Army's *Frontier Warfare, Small Wars* and *Internal Security*, as a response to the changing circumstances and the operational environments right after independence in 1947 till the promulgation of the SCW doctrine in 2013.

Pakistani Army's Inherited Counterinsurgency (*Small Wars*) Approach in 1947

The concept of institutional memory is very important for studying any organisational learning, but especially that of the military as Nagl maintains "the institutional memory of an organization is the conventional wisdom of an organization about how to perform its tasks and missions" (2005, p. 6). To trace the foundations of the institutional memory of the Pakistani Army with regards to fighting *small wars*, it is necessary to review the British Indian Army's lessons of the *small wars* especially those fought in the erstwhile Frontier Region of India (modern-day FATA, KPK, and Balochistan). It is essential because the Pakistani Army developed its doctrine and operational procedures after independence from British colonial rule, on this initially inherited firm foundation of the British Indian Army's legacy of fighting *small wars*, that is, Frontier Warfare, Operations in Aid to Civil Power and Imperial Policing.

Frontier Warfare

The no-mans-land in the mountainous region between the North West Frontier of India and Afghanistan was the site of many punitive operations and *small wars* by the British Indian Army against the intractable trans-frontier Pathan (also commonly known as Pashtun) tribes. For securing the NWFP, a force named the Punjab Irregular Force (PIF) was raised in the autumn of 1849 (Moreman, 1998, p. 6), re-designated as the Punjab Frontier Force (PFF) in 1865 (Moreman, 1998, p. 27) and later finally as Frontier Force (FF) (Moreman, 1998, p. 150, 174). At the time of the partition of India in 1947, all the regiments of FF were transferred to the Pakistani Army (Moreman, 1998, p. 185). These were the units which undertook operations against the tribal people after partition and later on they formed a major chunk of the initial onslaught of the Pakistani Army on terrorist hideouts in FATA in the GWOT after 9/11. The repeated military actions, ranging from minor skirmishes to large-scale campaigns involving tens of thousands of men (especially the FF and Gurkha regiments), against the trans-border Pathan (also commonly known as Pashtun) tribes in NWFP during the British Raj led to the development of a peculiar frontier-style of tribal policing and army operations against the tribal guerrillas called Frontier Warfare (Moreman, 1998, p. xxii).

Though the Frontier Warfare went through several phases of development in terms of the employment of troops, arms, weapons, and airpower, there were certain consistent characteristics, such as *"butcher and bolt"*, in the British Indian Army's approach to low intensity operations in the NWFP (Johnson, 2012, p. 218; Fair & Jones, 2010; p. 51). The army always resorted to heavy kinetic actions, seeking to deter through the presence of a large number of regular and irregular troops, campaigns to destroy the tribal *lashkars* (war parties) and burning or destroying their villages and crops. The fortified houses of the known tribal insurgent leaders were razed to the ground. Exactly the same measures were taken by the Pakistani Army in Balochistan (1973–78) and later during operations in FATA (2007–08). These will be discussed later in the chapter.

The Royal Air Force (RAF) was also employed in support of the ground forces during the inter-war years. The use of combat air support against insurgent guerrillas was established to be one of the effective principles of fighting mountain warfare. Therefore, a new joint military manual – the first of its kind – was published in November 1939, named the *Frontier Warfare-India (Army and RAF)*. All these operational procedures, tactics, and practices employed in the Frontier Warfare were in consonance with the principles of war, such as *Offensive Action, Maintenance of the objective, Concentration of Strong Force*, and *Use of Air*, enshrined in the Frontier Warfare Doctrine of the British Indian Army.

Military Aid to the Civil Power

Charles Gwynn in his famous and influential *Imperial Policing* manual (1934, cited in Johnson, 2012, p. 228) highlighted four governing principles of imperial policing: the "primacy of the civil power, the use of minimum force, firm and timely action, and, cooperation between civil and military authorities". The focus of Gwynn's argument was not the political concession to ward off political unrest but to take firm measures, which at times was based on collective punitive actions and coercion, for restoring law and order. This aggressive spirit was equally maintained in the soldier's pamphlet *Duties in Aid of the Civil Power* by stating "when soldiers are called upon to act, they act as soldiers and not as policemen" (1934, cited in Johnson, 2012, p. 228). *Duties in Aid of the Civil Power* (1934, cited in Johnson, 2012, p. 229) highlights the use of "necessary force" rather than minimum force as the fundamental and important principle in aid of the civil power (internal security operations). The Official Manual of the British Indian Army entitled *Duties in Aid of the Civil Power* highlighted six fundamental principles for internal security operations; the "provision of adequate force, necessity for offensive action, coordinated intelligence under military control, efficient intercommunication [between agencies], mobility and security measures to conceal preparation and movement of troops" (1937, cited in Johnson, 2012, p. 230). The manual (1937, cited in Johnson, 2012, p. 230) maintained its aggressive spirit by maintaining the provision of Martial Law during internal security duties and use of kinetic actions by highlighting that in case the rioters refused to disperse by verbal warnings the army would open fire not over the heads of the crowd but at them to inflict causalities. This was the case in the practice of the British Indian Army which did not restrain from the use of firepower on demonstrators. For example, during Gandhi's Quit India Movement (1942–44), the British Indian Army opened fire 369 times, resulting in the killing of 1000 individuals and wounding another 2000 (Mansergh, Lumby, & Moon, 1970, p. 933). The Pakistani Army continued this aggressive spirit by using heavy kinetic means against protestors, for example in Kalat State (Balochistan) in 1948. This led to the transformation of protest into the full-scale Baloch insurgency (first round).

The enduring lessons of the Frontier Warfare and the Operation in Aid of the Civil Power and the doctrine published in the pamphlets/manuals of the pre-partition period of India were mirrored in the counterinsurgency approach of the Pakistani Army for at least during the first decade after partition.

Evolution of Counterinsurgency Doctrine of the Pakistani Army

Kalat Uprising 1948

The newly formed Pakistani Army had to carry out its first operation in Balochistan in March 1948 when the princely State of Kalat refused to join the Federation of Pakistan and declared independence (Bennett-Jones, 2009, p. 132). As per the legal connotations, it was not strictly an operation in Aid to Civil Power because by then State of Kalat had maintained its independence status though the

Shahi Jirga (Quetta Municipality)[3] had voted in favour of Pakistan in 1947 (Ishtiaq, 2013, p. 67). The Pakistani Army took this operation initially as an Aid to Civil Power and adopted the same approach which was embedded in the legacy of the Imperial Policing of the British Indian Army. Because of the pressure of the Pakistani Army, the Khan of Kalat acceded to Pakistan on 27 March 1948 (Ishtiaq, 2013, p. 67).

In all the major towns like Pasni, Jiwani, and Turbat, demonstrations started against the forceful merger of the Independent State of Kalat to the Federation of Pakistan. There were two kinds of responses which the Pakistani Army had to face in the aftermath of the annexation, one was the mass agitation led by many of the leaders of the Kalat State National Party (KSNP) and the other through guerrilla war led by Prince Abdul Karim, the brother of the Khan of Kalat who was the then Governor of Makran (Ishtiaq, 2013, p. 67). Within a few weeks of this forceful annexation, Prince Abdul Karim fled to Afghanistan with his men.

By following its inherited military ethos, the Pakistani Army followed the principle of "necessary force" instead of minimum force. Excessive force was used to crush the peaceful agitation in the towns. Frequent curfews were observed, the ringleaders of the agitation were jailed, and the protestors were fired upon frequently. Although warning shots were fired initially to disperse the crowd, protestors were then directly fired upon to inflict casualties (Kamran 2018, Interview, 15 December). This was exactly the same tactic used by the British Indian Army during the Quit India campaign. The official record of casualties is not available (Kamran 2018, Interview, 15 December). The guerrilla leader Prince Abdul Karim was persuaded by the Pakistani Army through the Khan of Kalat to return back to Balochistan for negotiations with the Government of Pakistan. On his return on July 8, 1948, the prince along with his comrades was captured and imprisoned (Naseer, 2010, p. 526). Practically, the Pakistani Army had overcome the challenge of the guerrilla fight with Prince Abdul Karim relatively easily through persuasion and threats to the relatives of the comrades of the prince. The operation lasted for a few months.

Downie's Cycle

During this period Brigadier Akber Khan, the commander of the forces in Kalat, the war veteran of Waziristan Campaigns 1936–39, critically observed the development of operations and identified the two issues leading to the Army's performance gap. He argued that a clear interpretation and assessment of concepts of the "use of minimum force" and "use of necessary force" was required in codifying the Pakistani Army's approach to such operations. Secondly, the "rules of engagement", including the procedure of firing upon the protestors/rioters needed redefining. He endorsed his observation in the war diary of the Brigade HQ (Kamran 2018, Interview, 15 December). This corresponds to steps 1 and 2 in Downie's learning cycle, and the army's performance gap was identified. Brigadier Akber Khan began to think of alternative actions that might be more successful and collaterally less damaging (step 3 in the learning cycle where key military figures engage in search of better ways for fighting the *small wars*). Hence organisational

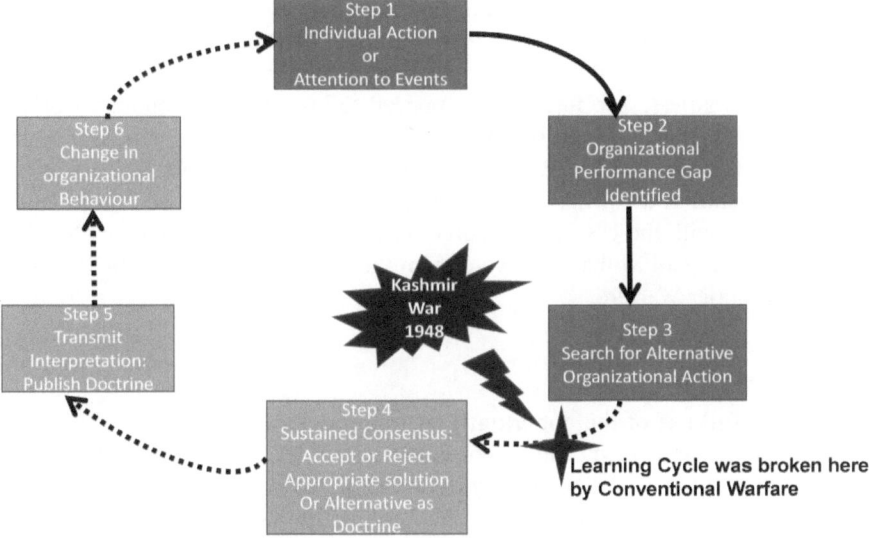

Figure 3.2 Pakistani Army's Learning Cycle (1948).

Source: Author.

learning was taking place with the learning cycle initiated and passing through its early stages. However, the process of learning became weak at step 3, and it could not progress beyond it. There was one big reason for the failure to progress beyond step 3, and that was the war with India in 1948. The army as an organisation could not get enough time to consolidate the solutions to the two problems which Brigadier Khan recognised during the operation (Kamran 2018, Interview, 15 December) as the conventional war between India and Pakistan (1948) claimed the attention of the whole of the army including Brigadier Khan (Ishtiaq, 2013, p. 74). Figure 3.2 (above) summarises this learning cycle.

Period from 1949–1965

At the time of the partition, there was a bitter dispute between Pakistan and India on the division of financial resources and military supplies (Fair, 2014, p. 56). Moreover, both countries were locked in a bitter rivalry over the territory of Kashmir (Fair, 2014, p. 57). This rivalry resulted in the first conventional war between Pakistan and India, just within a few months after the independence. Owing to the continuing Indian threat, the Pakistani Army remained mainly focused on developing conventional military doctrines during the first two decades of independence, 1947–1965. The only progress towards unconventional warfare was the formulation and publication of the Frontier Warfare GSP in 1963 and the raising of the Special Services Group (SSG) of the army, as the specialists of Frontier Warfare (later extensively used in the GWOT) (Muzammil, 2018, Interview, 27 December).

The Pakistani Army in 1956 established the elite SSG, a force of army guer-rillas, with the collaboration of the US Department of Defence and CIA (Nawaz, 2008). It was a stay-behind force in Pakistan in view of the Russian threat through the northern frontiers. The force was supposed to be expert in mountain warfare (Frontier Warfare) as it would be employed to fight the Russians should they in-vade via Afghanistan in NWFP. Lieutenant Colonel (Lt Col) A. O. Mitha was the first commandant of the SSG, and he was responsible for raising and the training the SSG along with the US Special Force officers in Cherat cantonment (NWFP) (Mitha, 2003). Lt Col Mitha was a veteran of the Burma campaign (WWII) and an expert on Frontier Warfare.

Downie's Cycle

Mitha made full use of the opportunity of training SSG, sheerly out of his pro-fessional interests, to study the recent operations (1952–53, 1954) of the Tochi Scouts (Frontier Corps (FC)) along with one FF battalion against the Faqir of Ipi in North Waziristan Agency (FATA). The Faqir of Ipi had taken up arms against the political administration of the agency and declared "Independent Pakhtuni-stan" (land of Pathans) with patronage from the Afghan Government. The FF unit and FC mainly followed the two principles of the *Maintenance of the objective* and *Offensive action* as enshrined in the Frontier Warfare pamphlets of the British Indian Army and well-practised by these units before partition. Mitha studied the operations through the War Diaries, Operation Completion Reports (OCRs) and the daily Situation Reports (SITREPS) of the FC/Army. Moreover, there were many personnel from the FF Battalion amongst the trainees of Mitha, who had recently participated in these operations against Ipi Faqir in Waziristan. This also helped him in getting insight into these operations of Frontier Warfare which were conducted for the very first time by the Pakistani Army/FC after independ-ence. Lt Col Mitha designed the SSG training on real-time scenarios based on the accounts of recent operations of the FC and an FF battalion against tribal Pathans (corresponds to step 1 of the learning cycle). He recognised the performance gap of the FC and FF units regarding excessive collateral damage caused by them due to "unnecessary trigger friendliness". He attributed this attitude to the legacy of the British Colonial Army. This corresponds to step 2 of the institutional learning process. He came up with the alternative approach of "Fire for Effect only" which meant that no unnecessary firing should be resorted to. Moreover, in his report about the Frontier Warfare to the General Staff (GS) Branch, GHQ he highlighted that there was a great need to sensitise the troops that the tribal people of Frontier and Balochistan were no more colonial subjects but the citizens of Pakistan and that the Government of Pakistan was responsible for their welfare (step 3 of the institutional learning cycle). The consensus developed and the GS branch, GHQ accepted the proposals in principle (step 4). As a result, in 1963 for the first time, the Pakistani Army published its GSP on Frontier Warfare incorporating Mitha's suggestions (step 5). Another factor which facilitated the step 5 (publication of Doctrine) of the institutional learning process for the Pakistani Army during this

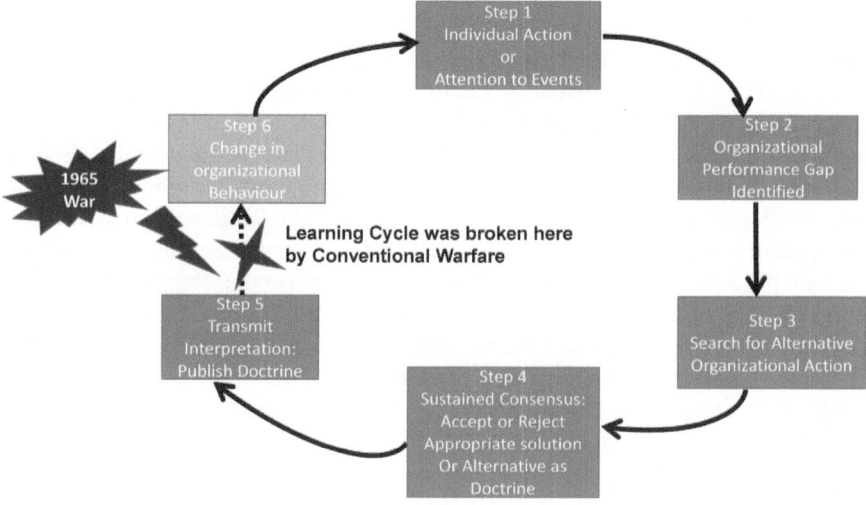

Figure 3.3 Pakistani Army's Learning Cycle (1949–1965).

Source: Author.

period was the establishment of the Research and Development (R&D) Directorate in 1959 at GHQ. The procedure for publication of GSPs was evolved in the Pakistani Army in 1963 through the R&D Directorate[4]. Though Mitha was successful in inculcating this new approach to the SSG the behaviour of the Pakistani Army, at large, could not be changed due to its mindset and focus on conventional warfighting with India during the early 1960s. Therefore, the last step of change in the organisational behaviour of the institutional learning cycle could not be achieved. Tensions escalated between the two arch-rivals, Pakistan and India, ultimately culminating in the war of 1965 between the two countries. Figure 3.3 (above) summarises this learning cycle.

Period from 1966–2000

During this period the Pakistani Army had to fight two major insurgencies; one in erstwhile East Pakistan (Bangladesh) in 1971 and the other in Balochistan (the fourth round of insurgency from 1973 to 1978). The former turned later into a full-scale war between Pakistan and India which Pakistan ultimately lost, resulting in the creation of independent Bangladesh. However, the Pakistani Army was successful in quelling the latter. The experiences/lessons learned during both these insurgencies, which were the most violent ones till then, left permanent marks on the counterinsurgency approach and culture of the Pakistani Army. This was later reflected in the SCW doctrine (2013). For example, the importance of the principle of "*Political Supremacy*" was learnt, through the hard way, and accepted as one of the fundamentals of counterinsurgency campaign although it was not practised until Operation Rah-e-Rast (Swat) in 2009.

East Pakistan Insurgency

During 1971, the Pakistani Army was used as a state instrument of power to quell the insurgency in East Pakistan by the Bengalis of East Pakistan for gaining separation from West Pakistan owing to the long deprivation and political grievances against unjust policies (Adeney, 2007). The insurgents formed a guerrilla resistance movement named Mukti Bahini (meaning freedom fighters) to fight a guerrilla war with the Pakistani Defence Forces in East Pakistan. The Pakistani Army conducted the operation, code-named *Searchlight,* and brutally suppressed the insurgency by the Mukti Bahini. The Mukti Bahini had full external support from India regarding organisational and operational planning for insurgent actions, recruitment, logistics, training, equipment, and funding. Despite the strong Indian support, the Mukti Bahini was not successful and therefore India on 4 December 1971 had to launch a full-scale conventional war against the Pakistani Army, after recognising the Bangladesh Government in exile on 3 December 1971. The war lasted 14 days (Gates & Roy, 2014a, p. 119). Pakistan lost the war primarily because of massive Indian intervention (Shafqat 2018, Interview, 24 December). Thus, East Pakistan turned into an independent country; Bangladesh.

Balochistan Operation 1973–1978

After the East Pakistan fiasco, in 1973, the insurgency surfaced in Balochistan as a consequence of the centralisation of powers by the Federal Government. The Federal Government attempted to threaten the Baloch identity and rights by dissolving the elected Provincial Government of Balochistan by the then PM of Pakistan, Mr. Zulfiqar Ali Bhutto on the charges of treason against the Federation. Mr. Bhutto established governor rule in the province of Balochistan, which introduced direct control of the central government over the province (Mahsood & Miankhel, 2013, p. 52). This stimulated the fourth round of insurgency (1973–78) in Balochistan. Mr. Bhutto, being the powerful Chief Martial Law Administrator as well, used military force extensively to counter the Baloch insurgency by all kinetic means (Nawaz, 2011, p. 3). The COIN operations by the Pakistani Army ousted most of the insurgents from the province to their hideouts in neighbouring Afghanistan. The COIN actions could not root out the insurgency completely.

The COIN forces crushed the insurgency with the help of Iranian 30 AH-1 Cobra attack helicopters. These attack helicopters neutralised the insurgent's edge of knowing the mountainous terrain of Balochistan better than the Pakistani Army (Heeg, 2012, p. 13). Moreover, the Pakistani Army bombarded and burned the encampments and crops of 15,000 Baloch families in Chamalang Valley with the aim of drawing back the insurgents from their mountainous hideouts to rescue the families (Heeg, 2012, p. 13). This tactic proved effective, and the insurgents were attacked heavily once they came down from the mountains. In short, the British Indian Army's legacy of the *"butcher and bolt"* was truly followed by the Pakistani Army in Balochistan.

The effective use of airpower was also one of the principles enshrined in the British Indian Army's Frontier Warfare Doctrine. Although by then the Pakistani Army had its own GSP on Frontier Warfare which highlighted the "least possible damage to life and property" that was not adhered to. Adeel (2019, Interview, 08 January) highlights that one of the reasons for still resorting to the old British Indian Army's legacy of *"butcher and bolt"* was that the new culture of inflicting the "least possible damage" could not be inculcated in the army (change in organisational behaviour, the step 6 of Downie's cycle, could not take place) because of the conventional war focus (less than two years after the publication of GSP on Frontier Warfare, Pakistan went into the war with India in 1965). In addition, the Pakistani Army was desperate to crush the insurgency by all available means to avoid another possible secession like Bangladesh in 1971 (Ahsan 2019, Interview, 16 January). Though the fourth round of the Balochistan Insurgency was an unambiguous COIN victory (Paul et al., 2013, p. 355) the root causes of the insurgency were not addressed. As per an estimate, about 80,000 Pakistani troops and 55,000 Baloch were engaged in the fighting at the peak of this conflict. The Pakistan military lost 300–400 personnel and somewhere between 7,300 to 9,000 Baloch insurgents were killed (Paul et al., 2013, p. 361).

Downie's Cycle

Step 1 (Attention to Events)

In the aftermath of losing the eastern wing of the country through an organised insurgency in 1971, a post-war committee was formed in GHQ under the direction of the Chief of General Staff (CGS) to study the reasons for the failure of the Pakistani Army especially during the insurgency part of the conflict. A post-conflict evaluation committee was also set up under the General Officer Commanding (GOC) 33 Division in December 1978 to identify and review the performance gaps of the Pakistani Army during the Balochistan conflict (Rahimuddin Khan[5] 2018, Interview, 16 December). The creation of these committees at the highest level shows that thorough attention was paid to the two events in the quest for improvement.

Step 2 (Organisational Performance Gap Identified) & Step 3 (Search for Alternative Organisational Actions)

The two committees consisted of senior officers involved in the respective conflicts. These committees carried out a thorough study of the events and took considerable time to prepare comprehensive reports explicitly identifying the strategic and tactical weaknesses in the counterinsurgency approach of the army during these conflicts. These committees recommended an alternative approach. The performance gaps identified by the two committees were, by and large, the same – clearly reflecting the organisational behaviour/approach of the army in counterinsurgency operations during that period.

The findings of these committees[6] along with the lesson learned from the COIN experience of FATA, Malakand, and Swat are indeed the basis on which the future counterinsurgency doctrine of the Pakistani Army is written. Lessons from the COIN experience in FATA, Malakand, and Swat will be discussed in detail later in the chapter. The above-mentioned two committees' findings will be cross-referred in the next chapter, where the SCW doctrine (2013) addresses them. Hence analysing these findings are important owing to their intrinsic nature to the SCW doctrine. The findings of the two committees, along with the analysis in the form of lessons learned are as follows.

(A) ABSENCE OF POLITICAL SUPREMACY

The concept of political supremacy was not accepted while planning and executing the counterinsurgency in East Pakistan as well as in Balochistan during the fourth round of insurgency (Rahimuddin Khan 2018, Interview, 16 December). In 1971, the political leadership was not taken into confidence before launching the crackdown operation against the Awami League and the Mukti Bahini in East Pakistan. The reason was that the operation was conducted during the Martial Law period imposed by General Yahya in 1969 and in 1971 he was the president of Pakistan. As a military dictator General Yahya Khan was not inclined to take the political leadership on board.

(B) LACK OF COMPREHENSIVE RESPONSE AND SOLE RELIANCE ON MILITARY

The sole reliance on military means was yet another performance gap in the overall counterinsurgency strategy in East Pakistan. Extensive, indiscriminate, and excessive use of the army to suppress a popular movement provided a spark to an already volatile situation. The use of force to solve a political problem or achieve a political objective is never durable. It is at best a temporary solution. Political and economic efforts to resolve the problem were missing. The failure to include political and economic prongs in the overall counterinsurgency strategy was highlighted in the Post-War Committee's Report on 1971 War in East Pakistan, which maintained that "Dependence on military element alone and excluding other elements [economic and political] led us to lose East Pakistan; certainly, it was not a rational approach" (Pakistani Army, 1972, p. 12). Finally, the reports asserted "the integrity of a nation does not depend on the strength and size of the military forces alone. Economic strength and political stability are as important if not more so for the security of a country" (Pakistani Army, 1972, p. 49).

In Balochistan, hardcore military operations were also solely relied upon. There was absolutely no attention paid to the political and economic dimensions of the issue until General Zia, after toppling the Government of Zulfiqar Bhutto, granted amnesty, in his new capacity as the President of Pakistan, to all the Baloch prisoners who were arrested by the latter's government on treason and armed violence charges during the fourth round of insurgency in Balochistan (1973–78) (Shaikh, 2014).

(C) ABSENCE OF CIVIL-MILITARY COORDINATION (CIMIC) AND INTER SERVICES
     COORDINATION

The absence of CIMIC and inter Services Coordination during the counterinsurgency campaign in East Pakistan was yet another factor which contributed to the defeat. As head of state and Commander-in-Chief, General Yahya Khan had gathered around him a select group of army officers, who were handling most of the affairs. For quite some time, civilians were also kept at bay: the bureaucracy was made to operate through two brigadiers, who denied Yahya the bureaucratic valuable advice on matters of national security (Pakistani Army, 1972, p. 81). This was the same attitude of the field formation commanders in East Pakistan, especially if they had to deal with Bengali civil servants. During this time the institutions meant to provide higher direction of war and inter services coordination, namely the Defence Committee of the Cabinet, National Security Council and Joint Services Secretariat laid dormant and hence inter services coordination could not be achieved.

The absence of Civil-Military Coordination, especially during the initial phases of the Balochistan operation, was also identified as a grey area which resulted in trouble for the Pakistani Army and ultimately favoured the insurgents for inflicting losses onto the army.

(D) NO MEDIA POLICY

The absence of a Media Policy for countering the foreign/Indian media and presenting facts to the international and domestic audience led to international political pressure and caused degradation of the morale of troops fighting the insurgency. The authorities in East Pakistan removed the international media rather than co-opting them. On the day of Operation Searchlight, foreign journalists in Dhaka were bundled into trucks and taken to the airport to be sent away immediately (Mostofa, 2017). This proved to be a blunder. The international media condemned the behaviour of the Pakistani Army in East Pakistan. In the same manner, an extreme media censorship strategy was adhered during the Balochistan conflict from 1973 to 1978.

(E) POOR INTELLIGENCE

Timely and accurate intelligence into the pre-insurgency scenario was identified to be another weak link. The correct and timely intelligence assessment of the internal turbulence in East Pakistan during the late 1960s and forthright presentation by intelligence agencies to the policymakers could have prevented a situation where the Bengalis resorted to seeking Indian assistance. Similarly, the absence of timely and accurate intelligence during the Balochistan conflict from 1973 to 1978 led to poor planning during the initial phases of the operation. This ultimately resulted in high numbers of casualties in the security forces.

For any insurgency to succeed, internal/external support is very important. Therefore, to fight insurgency, it is essential to find out the supporters of the insurgents in the local population and take every measure to eliminate internal support. But the army failed to eliminate the internal support of the insurgents during both the conflicts because of the lack of intelligence.

(G) CONVENTIONAL METHOD OF FIGHTING

One of the points highlighted in the reports was that conventional warfighting techniques against the hostiles may not succeed. The Post-Conflict Evaluation Committee's Report on Balochistan emphasised, "To defeat insurgents, Army has to adopt the same technique as adopted by hostile but in a better way" (Pakistani Army, 1979, p. 56). During the Balochistan operation, the troops learnt this hard lesson belatedly. In other words, this assertion was the realisation of the need to adapt to the counterinsurgency culture and approaches which best suit the insurgencies instead of existing inclination towards the conventional warfighting. This realisation and consensus at the top-level clearly correspond to step 4 of the Downie's learning cycle, which depicts consensus on appropriate solutions in the form of alternative doctrine.

Step 4 (Sustained Consensus: Accept or Reject Appropriate Solution or Alternative Doctrines)

The GHQ accepted the reports of the post-war committee and post-conflict committee, set up in 1971 and 1978 respectively, without any changes. Thus, sustained consensus was developed. This corresponds to step 4 of Downie's institutional learning cycle.

Step 5 (Transmit Interpretation: Publish Doctrine) & Step 6 (Change in Organisational Behaviour)

Although during the period from 1966 to 1978, a formal doctrine for counterinsurgency did not exist, the military leadership keenly articulated and circulated the lessons of the counterinsurgency campaigns in East Pakistan and Balochistan within the army. This means that organisational learning was taking place in the Pakistani Army with the learning cycle initiated and passing through the first four stages as mentioned above. However, at step 5 of the cycle, the learning process became weak as the formal doctrine was not published. Nonetheless, the reports of both the committees, set up for the post-evaluation of the conflict, were widely circulated within the Pakistani Army for inclusion into the formation training notes and biennial training cycles; hence the institutional learning cycle progressed until step 5. However, the progression from step 5 to step 6 (change in organisational behaviour) could not take place. The change in organisational behaviour comes through adopting the doctrine/approaches on which consensus is developed which in turn is possible through "training or broader military education for small wars" (Nolan, 2012, p. 47). Between 1966 and 2001, neither the specialised counterinsurgency training nor the education of the Pakistani Army was adapted in the light of the two committees' reports in the 1970s.

From 1966 to 1978, the India-centric focus and hence continuity in the development/refinement of conventional warfighting capability and doctrinal approaches was the main reason for the interruption of the learning cycle at this stage. However, after 1978, there were two major reasons, embedded in the post-1978

Pakistan-India regional security environment and the Cold War politics of the US-Pakistan alliance against the Soviets, for the interruption of the learning cycle at this stage.

The first reason was that from 1979 to 1988 under the rule of General Zia, the Pakistani military dictator and then president, the Pakistani Army's leadership was busy in fomenting Afghan "jihad" against the Soviets in Afghanistan at the behest of the CIA (Sial, 2013, p. 2). The second reason was that during the period of 1979–2001, there was a long series of events which led to conflict escalation between India and Pakistan. The famous Siachen conflict arose in 1984; followed by Operation *Brasstacks,* the military exercise by the Indian military in Rajasthan across the Pakistan Border (1986–1987). It was regarded as the largest ever troops mobilisation of the Indian Military. In response to *Brasstacks*, the Pakistani Army carried out a joint command field exercise named *Zarb–e–Momin* in 1989. In 1998 both the countries carried out successful nuclear tests. Between May and July 1999 India and Pakistan fought a conventional war in Kargil, a district of Kashmir. As Fair and Jones (2010, p. 44) put it, "… fighting 'counterinsurgency' had not been a major focus of the Pakistan Army; its training was largely geared toward a conventional war with India". Therefore, despite initiating and progressing the institutional learning cycle of counterinsurgency, the Pakistani Army could not reach step 6 of the cycle, that is, change in organisational behaviour. The conventional warfighting focus resulted in breaking this learning cycle. Moreover, during this time, the senior military leadership, like General Zia ul Haq, General Akhtar Abdur Rahman (Chairman Joint Chiefs of Staff Committee (CJCSC) of the Pakistani Armed Forces) and their successors, were traditionalists and believed that conventional warfare was the main role of the Pakistani Army (Zaki 2019, Interview, 20 January; Jaffer 2019, Interview, 17 January). This mindset also contributed to the breaking of the learning cycle at step 6. Figure 3.4 (below) summarises this learning cycle.

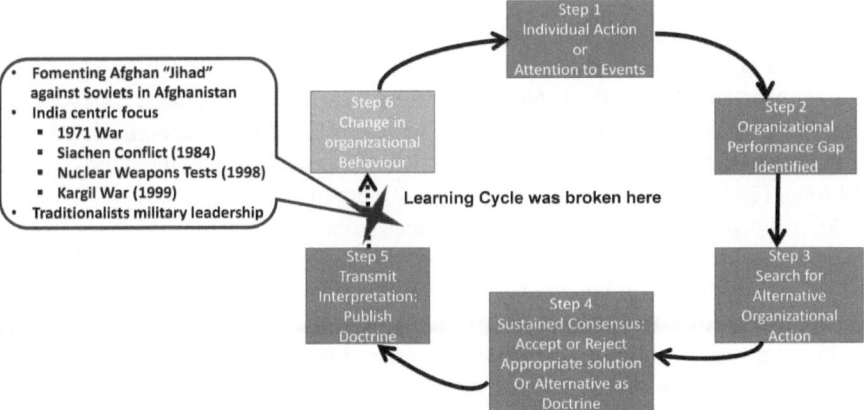

Figure 3.4 Pakistani Army's Learning Cycle (1966–2001).

Source: Author.

Period from September 11, 2001–2013

After 9/11, Pakistan joined the GWOT as a frontline state. From this point onwards till 2013, especially after 2006/7 when General Kayani took over as the Vice Chief of the Army Staff (VCOAS) and later as the Chief of the Army Staff (COAS), the Pakistani Army was fully focused on fighting the Jihadist insurgency in the FATA and NWFP then officially known as Low Intensity Conflict (LIC) within the army. Kayani played a great role in the Pakistani Army's organisational adaption, coordination, and in defining and embedding shared values and beliefs of counter-insurgency (Zaki 2019, Interview, 20 January; Adeel 2019, Interview, 08 January). His leadership style had a profound effect on the change in the Pakistani Army's organisational behaviour towards counterinsurgency and was a major influence on the organisational climate, defined by Nolan (2012, p. 21) as "the personality of an organisation". This resulted in the promulgation of first-ever SCW doctrine of the Pakistani Army in 2013, General Kayani's legacy.

The 9/11 attacks drew the US again into the region, after the cold war, but this time with the aim of ousting the Taliban and Al-Qaeda (Nawaz, 2011, p. 6). By then Pakistan had officially recognised the Taliban government in Kabul but owing to military and political pressure the then president of Pakistan General Pervez Musharraf assured full cooperation in the GWOT to the United States. Musharraf offered logistical support to the US forces during their Afghanistan invasion, in-cluding the establishment of US logistics bases in Balochistan (Gwadar, Pasni, Jacobabad, and Dalbandin). He also moved the Pakistani Army into the Pakistan-Afghanistan border areas (FATA) to clear it of insurgents and further sealing it so that the US forces could carry out their operations in Afghanistan effectively (Fair & Jones, 2010; Musharraf, 2006).

During these early operations, the Pakistani Army, except SSG, was neither prop-erly equipped for counterinsurgency nor trained for it. Fair and Jones highlight the fact that "Pakistan's security forces had limited experience in waging sustained op-erations in FATA before Operation Enduring Freedom, despite some experience in other areas" (2010, p. 44). The performance of the SSG was far better because they were good at Frontier Warfare owing to the legacy of Mitha's training regime for Frontier Warfare during the 1960s as mentioned earlier in the chapter. Despite all these drawbacks, the main objective of Operation Enduring Freedom, that is the overthrow of the Taliban Regime in Afghanistan, was achieved. In the words of Fair and Jones (2010, p. 44) "Pakistan played an important role".

After the successful clearing of tribal areas from Al Qaeda and Afghan Taliban insurgents by the army, a local uprising against the Pakistani Army started. There was a formative phase of insurgency in FATA between 2002 and 2007 (Rana, 2016). The Tehreek-e-Taliban Pakistan (TTP), led by Baitullah Mehsud, took arms against the Pakistani Army in 2006 and became the focal point of rebellion which spread rapidly. By 2009, it had developed into a full-scale insurgent movement, across FATA and the North West Frontier Province (now Khyber Pakhtunkhwa [KPK]) (Rana, 2016). Thus, all of a sudden Pakistan had to fight an insurgency which is sometimes referred to as "home-grown" although the army called it a "Low Inten-sity Conflict" (Nawaz, 2011, p. 3).

From 2001 to 2010, the army carried out many COIN operations in FATA, Swat, and Malakand regions. The major operations were (Nawaz, 2011, p. 8; Fair & Jones, 2010, p. 34):

- 2001–02: Operation Al Mizan (*The Balance*) in South Waziristan
- 2008: Operation Sherdil (*Lionheart*) in Bajaur
- 2008: Operation Zalzala (*Earthquake*) in South Waziristan
- 2008: Operation Rah-e-Haq (*The True Path/Faith*) in Malakand and Swat
- 2009: Operation Rah-e-Rast (*The Correct Path*) in Malakand and Swat
- 2010: Operation Rah-e-Nijaat (*The Path to Salvation*) in South Waziristan

The Pakistani Security Forces learnt a lot in fighting against the TTP and gradually improved their responses and strategies as Fair & Jones noted, "Pakistani operations had improved somewhat by Operation Sher Dil in 2008 (Bajaur), Operation Rah-e-Rast in 2009 (Swat), and Operation Rah-e-Nijat in 2009 and 2010 (South Waziristan). Frontier Corps and army forces were better able to clear territory and integrate operations with local tribes" (2010, p. xiv).

Downie's Cycle

In late 2010, the MO Directorate, GHQ, on the advice of the COAS, prepared the formal Lesson Learned notes of the counterinsurgency campaigns in FATA/NWFP from 2001 to 2010 (Nadeem 2018, Interview, 13 December). These were prepared in consultation with the MI Directorate, ISPR, and the various field formations involved in the operations. This effort corresponds to the first three stages of the Downie's institutional learning cycle, that is, Attention to Events, Identification of Performance Gaps and Search for Alternative Actions. The rest of the cycle for the above-mentioned period was completed in 2013, by additionally considering all the counterinsurgency campaigns of the Pakistani Army since 1947, as a result of General Kayani's endeavours for institutionalising the learning of the Pakistani Army. The remaining steps of the cycle will be discussed in the next chapter, under the heading *Kayani, Doctrine, Training and Institutionalisation of Learning*.

The lesson of these operations, in FATA/NWFP, in terms of the weaknesses and strengths of both the counterinsurgents and the insurgents had a long term effect on the development of the formal SCW doctrine and thus adopting a coherent counterinsurgency approach at army level.

Lessons of the COIN Experience: FATA, Malakand, and Swat

(a) Civil-Military Coordination (CIMIC)

One of the major issues which the Pakistani Army faced during the operations between 2001 to 2006 was the lack of civil and military coordination. During this time in FATA, no attempt was made to conceive a holistic approach to the battle against insurgents by cooperating with or including the local civil administration and the government departments. Even their superior knowledge of local physical and human terrain was not taken into consideration during the planning, executing or post-conflict phase (Nawaz, 2011, p. 9). The peculiar mindset of not taking the civil administration on board during the counterinsurgency operations accounts for the absence of initiatives like the National Solidarity Program (NSP) development which worked well in neighbouring Afghanistan (Nawaz, 2011, p. 14). Any such initiative would have helped in the rehabilitation of the internally displaced persons (IDPs) and the reconstruction of destroyed villages. From 2006 onwards, after the establishment of Civil Secretariat of FATA, the situation of CIMIC improved gradually against the backdrop of carrying out winning hearts and mind (WHAM) campaigns for the provision of infrastructure, building local capacities and ensuring sustainable livelihoods of the people in war-torn FATA (Kazmi, 2018, Interview, 19 December).

(b) Political Primacy/Ownership

During the initial years of GWOT, there was a trust deficit between the Pakistani Army and the political parties and populace. It was a one-sided decision of General Musharraf to be part of the global coalition against Taliban/Al Qaida. Under the dictatorial regime of Musharraf, the military actions in FATA and NWFP could not gain political support. People rejected the military option because of the widening unpopularity of Musharraf and also due to the great ambivalence possessed by the Pakistani masses about these military operations. The results of a survey of rural and urban Pakistanis conducted in October 2008 indicated that a majority of Pakistanis did not support the Pakistani Army fighting in FATA and NWFP (Fair & Jones, 2009, p. 179). The political ownership of the Pakistani counterinsurgency efforts in FATA and NWFP started to gather momentum after the return to civilian politics, once the trio of General Kayani, President Asif Ali Zardari and Prime Minister Yousaf Raza Gillani were at the helm after Musharraf stepped down in mid-August 2008 as the COAS and later as President. With the return of civilian supremacy, an era of greater demarcation of authority between civilian and military leadership occurred (Puri, 2012, p. 90). Although the legitimacy of this war remained controversial, the army's offensive actions in late 2008–2009 gained wider support from the government/political class and the populace than ever before (Puri, 2012, p. 91). In October 2008 for the first time, the President of Pakistan briefed both houses of the parliament, at a joint session, about the security situation. Earlier as a matter of practice, the parliament was always kept in the dark about the operations in FATA and NWFP. As per the US State Department cable:

> The October 8–9 closed joint session of parliament convoked by President Zardari to address the security situation has been widely praised as a good first step in convincing Pakistan's elected leadership to take ownership of the fight against extremism; this effort now should be extended to reach the general public. Director General of Military Operations Lieutenant General

Pasha's briefing to the group consisted largely of graphic video and other footage demonstrating what one parliamentarian called the inhumane and anti-Islamic nature of the militants ....

(Puri, 2012, p. 91)

The political ownership of the counterinsurgency campaign continued. For instance, in May 2009, the fourth round of military operations, named Rah-e-Rast ("The Right Way") was launched in Swat valley against the militants on the behest of the Provincial Government of the NWFP. This round of military operations, unlike the previous one, was unanimously backed in the National Assembly by the ruling Pakistan People's Party (PPP), the PML-N, PML-Q, the MQM, and the ANP (Puri, 2012, p. 96; Khadim, 2014). There was a strong political will behind the operation, which led to the success by clearing the Swat valley from the militants.

(c) Apparent Lack of Population Centricity and Use of Heavy Kinetic Means

The Pakistani Army, as in the past, focused a lot on heavy kinetic means against the insurgents while carrying out operations in FATA, Malakand, and Swat which often led to huge collateral damage. Fair and Jones (2010, p. 16) have identified this practice of the Pakistani Army as similar to that of the Russian army's approach in Grozny, Chechnya, where the latter crushed organised resistance for establishing control of the capital city and major towns through heavy use of artillery and air bombardments causing extensive bloodshed and damage to the cities. The Pakistani Army denies levelling the cities/towns and using lethal force against the civilian populace. However, one of the studies (2007) about FATA/Swat Operations at the Command and Staff College of the Pakistani Army identifies that the employment of decisive force against any target remains a very complex affair, as terrorist hideouts are mostly within built-up areas. The use of full force [including medium and heavy artillery and air bombardments] against the terrorists resulted in collateral damage. This proved to be counterproductive (Pakistani Army, 2007, p. 60). Fair and Jones (2010, pp. xiii, 82) attributed the persistent threat of militants in FATA/NWFP as caused by the failure to develop an effective population-centric counterinsurgency strategy in line with the recent innovations in the COIN warfare.

The Pakistani Army does not agree that it failed to adopt a population-centric approach during the operations in FATA/NWFP. As an initiative in line with the population-centric approach, the army claims it assisted the local inhabitants to leave the areas of its operation so that minimum loss to civilian life occurred; though it sometimes fostered unnecessary resentment in the people. The local inhabitants before every operation since 2008 were shifted to the Internally Displaced Persons (IDPs) camps established by the army/federal government at the settled areas of NWFP like Peshawar, Tank, Kohat, Bannu and D.I. Khan (Ahsan 2019, Interview, 16 January).

(d) Information/Psychological Operations

Lack of effective Information and Psychological Operations (or "psywar") was another weak and often missing link, as in the past, during operations in FATA/ NWFP. Nawaz highlights:

> Even the psychological elements of military planning seemed to have been forgotten. The only persons involved with psy-warfare were officers assigned to the Inter Services Public Relations (ISPR) Directorate, the media arm of the military, most of whom had no formal training in these tasks.
>
> (2011, p. 9)

In contrast, the insurgents had an effective media campaign in FATA and NWFP, targeting the local population in particular, and the general public at large. They professionally produced and marketed CDs dubbed into Pashto language highlighting the religious legitimacy of their campaigns against the army. They urged the local population to take up arms against the Pakistani Army as a Jihad while portraying army personnel as the friends of the "infidels" (Bush and Blair). The use of psywar by the insurgents was not limited to the CDs, but they also effectively used the FM pirate radio broadcasts to further their propaganda (Nawaz, 2011, p. 23). For instance, Mullah Fazlullah, leader of the TTP, continued broadcasts through his mobile FM radio station during operations in Swat (Fair & Jones, 2010). He was also known as *Mullah Radio* amongst the locals for his effective use of FM. Towards the end of 2009, the Pakistani Army paid attention to the information and psy-warfare, but its execution of it was poor. A seminar presentation (2008), conducted at the Pakistani Army Divisional Headquarters, reveals that the need for psy-warfare in FATA was highlighted because of the misconception of Jihad in the minds of people in FATA as a result of lack of awareness, no media access and continuous brainwashing and propaganda by Taliban and extremists. An effective psywar campaign could have broken the allegiance of the local people from the local Mullahs who were strengthening hostile public opinion against the GWOT (Pakistani Army, 2008).

(e) Intelligence

During the initial period of the GWOT, the intelligence was mainly supplied to the Pakistani Army for planning operations by the US forces through the Technical Monitoring Centres (TMCs) established in Miranshah (North Waziristan) and Wana (South Waziristan). ISI and MI directorates were in close liaison with the TMCs. The basic flaws observed, in human intelligence, initially had been the lack of focus, the lack of integration, and the inability of the intelligence operatives to penetrate the local insurgent network.

Too much reliance on the technical intelligence information provided by the US forces, lack of equipment and capabilities of the Pakistani Army and later closure of TMCs by the US resulted in grey areas as to the correct, effective and real-time intelligence information about the insurgents. However, the situation improved remarkably from 2007 onwards after the Pakistani Army procured Luna X 2000 German unmanned aerial vehicles (UAVs) in service with the Bundeswehr (German Army).

It enhanced the close reconnaissance capacity (over the hill, up to 65 km away) of the army considerably. Moreover, the UAVs were used for transmitting live video data, taking higher resolution still images, and Electronic Countermeasures (radio/radar jamming). The Pakistani Army also improved human intelligence capabilities to a greater degree over the years. For instance, in 2009 during the Operation Rah-e-Rast, the field formation had very reliable and actionable intelligence to plan and execute the military operations owing to the enhanced intelligence gathering capabilities of the Pakistani Army and no more dependence on the US regarding it (Muzammil, 2018, Interview, 27 December; Adeel 2019, Interview, 08 January).

(f) Training

As far as the training of the troops is concerned, the Pakistani Army initially lacked expertise in Frontier warfare except for the SSG units (trained under Mitha in the 1960s) which proved to be the best; followed by (to a lesser degree) the traditional FF units raised as the British Indian Army units before the partition of India. These two military establishments were carrying their traditional military ethos and culture of fighting frontier warfare. They were extensively employed in operations throughout FATA/NWFP, frequently recycled and rotated for a series of operations till 2008 as compared to other infantry units which were mainly raised after the partition, did not have Pathans (also commonly known as Pashtuns) in their ranks, and they possessed no prior experience of frontier warfare.

Conclusion

The weaknesses observed in the form of performance gaps of the army and the lessons learned of the various counterinsurgency campaigns since 1947 show the blueprint of the Pakistani Army's institutional learning over the period of roughly six decades set over the foundations of ethos and culture of fighting the *small wars* inherited from the British Indian Army in 1947. There were broadly four learning cycles and out of these, only the last one reached the terminal stage that is the change in the organisational behaviour (Step 6) as shown in Figure 4.1 in the next chapter. There were several structural issues which impeded the institutional learning process in various learning cycles of the Pakistani Army. The most important underlying structural issue was the frequent interstate conventional wars between Pakistan and India and linked to it was the traditionalist mindset of the senior military leaders who believed that conventional war with India was the primary role for the Pakistani Army (which the 1948, 1965, 1971, and 1999 wars seemed to vindicate). Besides this, there were certain other structural factors which impeded the learning such as the geostrategic priorities of the Pakistani Army during the Cold War era and later in the GWOT as discussed earlier in the chapter.

The lessons learned out of all the learning cycles are similar. Apart from other voids in the COIN strategy, the most fundamental of all, the political component – the primacy of political aim for counterinsurgency – was missing throughout in all the three cycles and initially missing in the fourth learning cycle (until General Musharraf stepped down as COAS). This can be attributed to the fact that most of the insurgencies were fought under the dictatorial military regimes in Pakistan. The country was under intermittent military rule for over 30 years. The Pakistani Army learned and adopted the principle of political primacy in the fourth cycle, and later it was reflected in the SCW doctrine. This aspect will be dwelled upon in the next chapter.

The enduring lessons learned in various cycles proved to be the cornerstone of the SCW doctrine promulgated in 2013 on which the Pakistani Army claims to have designed its current counterinsurgency strategy in Balochistan to provide safety to the CPEC (Kamran 2018, Interview, 15 December; Muzammil, 2018, Interview, 27 December). Hence these lessons learned are very important. The next chapter will look at the efforts for institutionalisation of this learning of the Pakistani Army which ultimately resulted in the *Change in Organisational Behaviour;* the changed mindset from a conventional war focus to the acceptance of the fact that COIN demands a different approach than "butcher and bolt" and specialised training in the light of SCW doctrine (2013) (Step 6 of the Downie's Cycle). The themes enshrined in the SCW doctrine (2013) will also be analysed in Chapter 4.

Notes

1 An infamous insurgent against whom the British Indian Army fought campaigns in FATA in the late 1930s, declared the 'Independent Pakhtunistan' (land of Pathans).
2 General Staff Publication (GSP) is any tactical, technical, training, or operational manual published by the Pakistani Army's Inspector General Training and Evaluation (IGT&E) branch, General Headquarters (GHQ), Rawalpindi. It is equivalent to US field manuals.
3 At the time of their departure from Sub Continent, the British conducted a referendum to decide the future of British Balochistan (the northern areas of Balochistan including Bolan Pass, Quetta, Nushki, and Naseerabad). In the referendum the Shahi Jirga (excluding the representatives of the Balochistan States) and the elected members of the Quetta Municipality (except its ex-officio members) participated within the electoral college. The Shahi Jirga unanimously opted for the merger of the province with Pakistan (Talbot, 1989, p. 119; Kausar & Inamul Haq, 1993; Waseem, 2011).
4 During the early years of independence, Pakistan Army had no training publication of its own. British Army pamphlets and pre-independence Indian Army Publications were in use till early 1960s (Muzammil, 2018, Interview, 27 December).
5 Rahimuddin Khan (late) was a retired four-star general of the Pakistani Army. He served as the Chairman Joint Chiefs of Staff Committee (CJCSC), which is, in principle, the country's highest-ranking and senior-most uniformed military officer. He also served as the 7th governor of Balochistan Province from 1978–1984 (Saeed, 2022). Rahimuddin Khan assumed the governorship at the final stages of the fourth round of the Balochistan insurgency (1973–78). After being appointed the Governor while still on active military service, on 16 September 1978, he immediately declared an end to the ongoing counterinsurgency operation. Rahimuddin Khan compelled the then President of Pakistan, General Zia ul Haq (the military dictator), to announce a general amnesty for Baloch fighters willing to give up arms. He died on 22 August 2022.

6 Rahimuddin Khan (late) presented the author with printed reports from these committees, which he retrieved from his 20th-century military campaign trunk made of galvanized steel (painted black) that bore his army number and name in white (hand painted). The trunk was a legacy of the British Army, and such trunks are still in vogue in Pakistani Army. The reports were inside an old A4-sized brown cardboard folder with the Pakistani Army logo printed at the top. The reports were manually typed, and the pages were yellow and embrittled.

References

Adeney, K. (2007). 'Democracy and Federalism in Pakistan', in He, Galligan, B. & Inoguchi, B. (eds.) *Federalism in Asia*. Cheltenham: Edward Elgar Publishing Ltd, pp. 101–123.

Adeel, K. (2015). 'Renewed Ethno Nationalist Insurgency in Balochistan, Pakistan: The Militarized State and Continuing Economic Deprivation', in Chima, J. S. (ed.) *Ethnic Sub Nationalist Insurgencies in South Asia: Identities, Interests and Challenges to State Authority*. Abingdon, Oxon: Routledge, pp. 124–142.

Bennett-Jones, O. (2009). *Pakistan: Eye of the Storm* (3rd ed.). New Haven, CT: Yale University Press.

Bono, G. (2004). 'The EU's Military Doctrine: an Assessment', *International Peacekeeping*, 11 (3), pp. 439–456. doi: 10.1080/1353331042000249037.

Cassidy, R. M. (2006). *Counterinsurgency and the Global War on Terror: Military Culture and Irregular War*. New York, NY: Praeger.

Dewar, J., August, D. & Builder, C. (1996). *Army Culture and Planning in a Time of Great Change*. Santa Monica, CA: RAND Publishers.

Downie, R. D. (1998). *Learning from Conflict: The U.S. Military in Vietnam, El Salvador, and the Drug War*. New York, NY: Frederick A. Praeger.

Fair, C. C. (2014). *Fighting to the End: The Pakistan Army's Way of War*. New York, USA: Oxford University Press.

Fair, C. C. & Jones, S. G. (2009). 'Pakistan's War Within', *Survival*, 51 (6), pp. 161–188. doi: 10.1080/00396330903465204.

Fair, C. C. & Jones, S. G. (2010). *Counterinsurgency in Pakistan*. Santa Monica, CA: RAND Corporation.

Fuller, J. F. C. (1926). *The Foundations of the Science of War*. Reprint, Fort Leavenworth, KS: U.S. Army Command and General Staff College Press, 1993.

Gates, S. & Roy, K. (2014a). *Unconventional Warfare in South Asia: Shadow Warriors and Counterinsurgency*. Abingdon, Oxon: Routledge.

Heeg, J. (2012) *Insurgency in Baluchistan*. Fort Leavenworth, KS: Kansas State University Press. Available at: https://baluchsarmachar.files.wordpress.com/2012/07/insugency-in-balochistan_final.pdf (Accessed: 15 March 2022).

Ishtiaq, A. (2013). *Pakistan Garrison State: Origins, Evolution, Consequences (1947–2011)*. Karachi, Pakistan: Oxford University Press, Pakistan.

Johnson, R. (2012). 'Small Wars and Internal Security: The Army in India, 1936–1946', in Alan, J. & Rose, P. (eds.) *The Indian Army, 1939–47*. Farnham: Ashgate publishing limited, pp. 215–220.

Kausar, I. (1993). 'Pakistan Resolution and Balochistan', *Journal of the Pakistan Historical Society*, 41 (4), pp. 393–394.

Khadim, H. (2014) 'Lessons from Swat', *Dawn,* 22 July. Available at https://www.dawn.com/news/1120694 (Accessed: 23 September 2022).

Lange, P. H. (1992). 'Understanding Military Doctrine', in Valki, L. (ed.) *Changing Threat Perception and Military Doctrines*. London: Palgrave Macmillan, pp. 1–17.

Long, A. (2016). *The Soul of Armies: Counterinsurgency Doctrine and Military Culture in US and UK*. Ithaca, NY: Cornell University Press.

Mäder, M. (2004). *In Pursuit of Conceptual Excellence: The Evolution of British Military-Strategic Doctrine in the Post-Cold War Era, 1989–2002*. Bern: Verlag Peter Lang.

Mahsood, A. & Miankhel, A. K. (2013). 'Baluchistan Insurgency: Dynamics and Implications', *Global Advanced Research Journal of Social Science (GARJSS)*, 2 (3) (March), pp. 51–57.

Mansergh, N., Lumby & Moon, P. (1970). *Constitutional Relations between Britain and India: The Transfer of Power 1942–47, vol. II*. London: Allen and Unwin.

Mitha, A. O. (2003). *Unlikely Beginnings*. Karachi: Oxford University Press.

Moreman, T. (1998). *The Army in India and the Development of Frontier Warfare, 1849–1947*. London, England: Palgrave Macmillan.

Mostofa, A. (2017) 'Politics and the press during 1971', *Dhaka Tribune*, 23 August. Available at https://www.dhakatribune.com/tribune-supplements/world-tribune/2017/03/26/politics-press-1971/ (Accessed: 25 March 2022).

Musharraf, P. (2006). *In the Line of Fire*. New York, NY: Simon & Schuster Ltd.

Nagl, J. (2005). *Learning to Eat Soup with Knife*. Chicago, IL: University of Chicago Press.

Naseer, G. K. (2010). *Tareekh-e-Balochistan (History of Balochistan)* (5th ed.). Quetta: Kalat Publisher.

Nawaz, S. (2008). *Crossed Swords: Pakistan, its Army, and the Wars Within*. Karachi: Oxford University Press.

Nawaz, S. (2011). *Learning by Doing – The Pakistan Army's Experience with Counterinsurgency*. Washington, DC: The Atlantic Council.

Nolan, V. (2012). *Military Leadership and Counterinsurgency: The British Army and Small War Strategy Since World War II*. London: I.B.Tauris.

Pakistani Army (1972). *Post-War Committee's Report on 1971 War in East Pakistan*. Rawalpindi: Chief of General Staff (CGS) Secretariat, GHQ.

Pakistani Army (1979). *Post-Conflict Evaluation Committee's Report on Balochistan-1979*. Rawalpindi: Chief of General Staff (CGS) Secretariat, GHQ.

Pakistani Army (2007). *Dynamics of FATA and Operation Al Mizan*, Group Research Paper. Quetta: Command & Staff College.

Pakistani Army (2008). *Psychological Operations in FATA* [Seminar Presentation]. Peshawar: HQ 7 Division.

Pakistani Army. (1990). *Glossary of Military Terms and Definitions*. General Headquarters (GHQ), Rawalpindi, Pakistan.

Pakistani Army. (2013). *Sub Conventional Warfare (SCW) Doctrine (Publication No. AP 2601E)*. Rawalpindi: Doctrine & Evaluation Directorate, GHQ. Available at: https://www.scribd.com/document/474069737/Sub-Conventional-Warfare-Doctrine-pdf (10 August 2022)

Paul, C., Clarke, C. P., Grill, B. & Dunigam, M. (2013). *Paths to Victory, Detailed Insurgency Case Studies*. Santa Monica, CA: National Defence Research Institute, RAND Corporation.

Puri, S. (2012). *Pakistan's War on Terrorism*. Abingdon, Oxon: Routledge.

Rana, M. A. (2016) 'Our counterinsurgency doctrine', *Dawn*, 04 December. Available at: https://www.dawn.com/news/1300320 (Accessed: 25 March 2022).

Sabharwal, S. (2022). *India's Pakistan Conundrum: Managing a Complex Relationship*. Abingdon, Oxon: Routledge.

Saeed, A. (2022) 'From military to politics', *The News on Sunday*, 04 September. Available at https://www.thenews.com.pk/tns/detail/988041-from-military-to-politics (Accessed: 15 September 2022).

Shaikh, A. (2014) 'A leaf from history: Reclaiming Balochistan, peacefully', *Dawn,* 05 October. Available at: https://www.dawn.com/news/1135570 (Accessed: 20 February 2022)

Shultz, H. Jr. (2012). *Organizational Learning and the Marine Corps: The Counterinsurgency Campaign in Iraq.* Newport, RI: Center on Irregular Warfare & Armed Groups (CIWAG) US Naval War College.

Sial, S. (2013) *Pakistan's Role and Strategic Priorities in Afghanistan since 1980.* The Norwegian Peacebuilding Resource Centre (NOREF). Available at: https://www.files.ethz.ch/isn/165432/9bc5b02e91c5a9b8ba49a5c46dbfd41a.pdf (Accessed: 15 April 2022).

Spencer, J. (2016). 'What is Army Doctrine', *Modern War Institute at West Point,* 21 March. Available at https://mwi.usma.edu/what-is-army-doctrine/ (Accessed: 22 September 2022).

Talbot, I. (1989). *Provincial Politics and the Pakistan Movement: The Growth of the Muslim League in North-West and North-East India 1937–47.* Oxford University Press.

UK. Army (2011). *Army Doctrine Primer (AC 71954).* Available at: https://assets.publishing.service.gov.uk/government/uploads/system/uploads/attachment_data/file/33693/20110519ADP_Army_Doctrine_Primerpdf.pdf (Accessed: 05 March 2022).

US Department of Defence. (2016). *Dictionary of Military and Associated Terms,* Joint Publication (JP) 1–02. Available at https://fas.org/irp/doddir/dod/jp1_02.pdf (Accessed: 24 February 2022).

Waseem, A. (2011) 'Balochistan: Accession at Gunpoint'*, Europe Solidarie* Available at: https://www.europe-solidaire.org/spip.php?article19762 (Accessed: 18 July 2022).

Wilkens, A. (2015) 'The Crowded-Out Conflict: Pakistan's Balochistan in its fifth round of insurgency', *Afghanistan Analysts Network,* 16 November. Available at https://www.afghanistan-analysts.org/the-crowded-out-conflict-pakistans-balochistan-in-its-fifth-round-of-insurgency/ (Accessed: 15 March 2022).

4 Institutionalisation of Learning and the Themes of the Sub-Conventional Warfare Doctrine

Introduction

The current chapter analyses the institutionalisation of the learning of the Pakistani Army with respect to counterinsurgency, as thoroughly discussed in the last chapter, since independence in 1947. This learning led to the promulgation of the Sub-Conventional Warfare (SCW) doctrine of the Pakistani Army for the very first time in 2013. The institutionalisation of the learning is a very crucial aspect for the organisation on the whole as Catignani reiterates:

> Without the incorporation and institutionalization of new knowledge within the organization, such knowledge is lost once personnel (or units) have moved on or ceased to exist within the organization ...... the institutionalization of new knowledge [is] a key indicator of the extent to which the Army has realized organizational learning.
>
> (2014, p. 32)

Through the implementation of various training regimes, the institutional learning cycle for the very first time reached the last step of Downie's learning model, that is, the *change in the organisational behaviour* of the Pakistani Army. The SCW doctrine (2013) is the essence of the counterinsurgency experience of the Pakistani Army based over six decades – transformed and innovated as per the modern requirements of the army through a systematic and analytical process. The doctrine sets out formally the current counterinsurgency approach of the Pakistani Army (Ahsan 2019, Interview, 16 January).

This chapter firstly discusses the institutionalisation process of the learning, initiated by General Kiyani and the efforts and stimuli to reach the consensus of promulgating the SCW doctrine. This is followed by tracing the change in the *organisational behaviour* of the Pakistani Army through the training regimes. Thereafter, the chapter looks at the themes enshrined in the SCW doctrine (2013) and how it addresses the performance gaps of the army highlighted in Chapter 3. It is essential to highlight the themes expressed by the doctrine because "Army doctrine provides a common language and a common understanding of how Army forces conduct operations" (US Army FM 3–0 *Operations*, 2001 cited in Nagl, 2005, p. 7).

DOI: 10.4324/9781003413905-4

The chapter also explicitly explains the rationale of discussing various themes of the SCW doctrine (2013) and their usage in the later empirical chapters.

Kiyani, Doctrine, Training, and the Institutionalisation of Learning

Proximate Causes of the Promulgation of a Formal Doctrine in 2013

In 2011, after being confirmed as Chiefs of the Army Staff (COAS) for a further three years in July 2010 (MacDonald, 2010), General Kiyani focused on conceptualising the intellectual aspects of Pakistani COIN. This was for two main reasons. From 2010, the Pakistani Army continued its military operations in the tribal areas. Although the insurgents retaliated with devastating terrorist attacks, their focus was not on occupying territory, unlike in the past, but on conducting terrorist attacks (Puri, 2012, p. 105). This situation posed a far less strategic threat to Pakistan compared to the recent past (Lieven, 2011, p. 476) and provided some respite to the Pakistani Army in the operational areas. This provided an opportunity for the top brass to ponder over the intellectual aspects of the COIN (Asim 2019, Interview, 25 January).

The second reason was US pressure to adopt a formal counterinsurgency doctrine. The US was greatly concerned over the non-existence of such doctrine. For example, Fair & Jones noted that "Pakistan's lack of an official counterinsurgency doctrine remains a lingering challenge" (2010, p. xiv). After General Ashfaq Kiyani's appointment as COAS (end of 2007), US officials were hopeful that the Pakistani Army would devise and formally adopt the COIN strategy. However, Kiyani reiterated that the army's priorities were to exist as a conventional force poised towards India rather than becoming a counterinsurgency force (Rashid, 2008). US officials and commentators believed that between 2008 and 2009, Pakistan had an uneven commitment to GWOT as its main focus on conventional warfare with India (Fair & Jones, 2009, p. 162). The Pakistani Army used the term Low Intensity Conflict (LIC) for the counterinsurgency operations in FATA [until 2010] (Nawaz, 2011, p. 9) as opposed to the US preferred term of counterinsurgency. The difference in the terminology was significant and reflected the major bias of the Pakistani Army to conventional warfighting. Nawaz (2011, p. 4) highlights that in 2008 the Director General (DG) Military Operations (MO), then Major General Pasha, articulated the concept of fighting insurgency by saying, "all you need is a well-trained infantry soldier" – a view that was more in line with the conventional warfighting approach. From 2010, however, Pakistani Army officers realised a broader approach was needed to fight an irregular war within their territory (Nawaz, 2011). The Pakistani Military officials increasingly adopted the term counterinsurgency in bilateral forums in place of LIC because of the US focus on the concept. Nevertheless, the operational concepts were in line with the LIC despite this change in the use of the terminology (Fair & Jones, 2010). Against the backdrop of this prevailing scenario, General Kiyani decided to develop the SCW doctrine of the Pakistani Army, which would establish the basic contours of the irregular warfare approach of the army for any future conflict. The approach enshrined in the SCW doctrine was adopted by the security forces in Balochistan after 2013 (after the CPEC agreement) (Asim 2019, Interview, 25 January; Nadeem 2018, Interview, 13 December; Adnan 2019, Interview, 14 January). The formal approach to counterinsurgency in the form of SCW doctrine (2013) will be discussed in the next section whereas the practical manifestation of the counterinsurgency approach and its efficacy in the context of Balochistan will be evaluated later in Chapter 6 of the book.

The first step taken by General Kiyani was the revamping of the D&E Directorate at GHQ to formulate the doctrine. Formerly, the focus of the directorate was publishing the GSP as per the scripts provided by the concerned arms' directorates; there was virtually no intellectual input by the D&E in the publications (Muzammil, 2018, Interview, 27 December; Jaffer 2019, Interview, 17 January). Kiyani furnished the D&E directorate with the requisite intellectual toolkit by posting officers to it with better professional intellect/qualifications, mostly the graduates of the Command & Staff College (C&SC), National Defence University (NDU), and the Royal College of Defence Studies (RCDS), UK (Muzammil, 2018, Interview, 27 December; Jaffer 2019, Interview, 17 January). Moreover, additional funds were allocated to the Directorate to equip it with the requisite IT facilities for research. Another significant step was that two additional roles were assigned to the D&E directorate, to (1) develop and promulgate concepts and doctrines for the army, adapting to current, and future challenges, (2) act as the army's think tank to provide input on hardcore professional matters. These measures played a major role in transmitting and embedding the Pakistani Army's approach to counterinsurgency (Kazmi, 2018, Interview, 19 December; Muzammil, 2018, Interview, 27 December).

Downie's Cycle (the Period from 9/11–2013)

The first three steps of the cycle, for the period after 9/11 till 2013, were completed in 2010 in the form of the Lesson Learned notes drafted by the MO Directorate, as discussed in the last chapter. The rest of the three steps of the cycle for the period mentioned above were completed as a consequence of General Kiyani's holistic approach for the institutionalisation of the learning in the Pakistani Army. On his advice, the D&E Directorate considered the Lesson Learned notes for the period after 9/11 in developing the SCW doctrine of the Army. It also carried out thorough scrutiny of operational plans and post-operation reports of all the internal security operations, operations in aid to the civil power and the counterinsurgency operations including the ongoing operations in FATA conducted by the Pakistani Army since 1947. Moreover, the D&E Directorate thoroughly consulted the Post-War Committee's Report on 1971 War in East Pakistan (Pakistani Army, 1972) and the Post-Conflict Evaluation Committee's Report on Balochistan (Pakistani Army, 1979). It also interviewed the officers (mainly retired) who planned/participated in these operations (Rahimuddin Khan 2018, Interview, 16 December). This enabled the D&E Directorate to evaluate systematically the operational, strategic, and tactical strengths and weaknesses of these campaigns. Several formal In-House

Discussions (IHDs) were arranged with concerned directorates such as MO Directorate, Military Training (MT) Directorate, Military Intelligence (MI) Directorate, all the fighting and supporting arms directorates, 11 Corps (based in Peshawar, KPK), and the Southern Command (based in Quetta, Balochistan). The input was also sought from various training establishments such as the School of Infantry & Tactics (SI&T), Quetta; Pakistan Military Academy (PMA), Abbottabad; Command & Staff College (C&SC), Quetta and Special Services Group (SSG) Training Centre, Chirat (KPK). The war diaries of all the FF regiments raised as part of the British Indian Army who participated in various campaigns in FATA/NWFP before and after the partition of India were also thoroughly studied. This was particularly helpful in looking at the historical dimensions of the operations in the frontier region by the FF units and their operational culture build-up over the firm foundations set by the British Indian Army. That is how the current COIN culture has its historical roots very much embedded in the British Indian Army's culture of fighting *small wars*. The D&E Directorate took considerable time in carrying out this massive task. During the process of developing the SCW doctrine, General Kiyani personally chaired many IHDs out of his desire to search the alternatives against the backdrop of performance gaps identified. The IHDs were held to develop consensus at the GHQ level about various aspects of the proposed SCW doctrine. This corresponds to step 4 of Downie's Cycle. Finally, in 2013, the draft of the SCW doctrine was approved by the COAS, and it was promulgated, for the very first time, by the army (The Express Tribune, 2013). The promulgation of the doctrine corresponds to step 5 of Downie's institutional learning cycle.

The Step 6 of the cycle, that is *Change in the Organizational Behaviour*, took place after the retirement of Kiyani but was implemented through his legacy of COIN training and teaching regimes of the Pakistani Army designed during his command and put in place during the tenure of his immediate successor, General Raheel Sharif. After the promulgation of the SCW doctrine (2013), the Pakistani Army started taking the subject of COIN seriously and formalised COIN training. This was a significant change in the organisational behaviour of the army. Formal COIN training was included in the curricula for the first time, in the form of seminars, lectures, workshops, scenario-based training, Tactical Exercises Without Troops (TEWTs) – a British Army legacy – and War Games, at the premier training institutions of the Pakistani Army namely, the SI&T, PMA, C&SC, and NDU. Various training initiatives, such as *Army Training Year, Army Firing Year,* and the *Year of Soldiers*, were also introduced by Kiyani as an endeavour to inculcate the new COIN spirit throughout the Pakistani Army. The rigorous Pre-Induction Training (PIT) Cycle was introduced in line with the SCW doctrine. The PIT cycle, of 16 weeks, is conducted in two equal phases of all the units going to be employed in the counterinsurgency operations. The training is formally conducted under the arrangements of the local formations and Inspector General Training and Evaluation (IGT&E) branch, GHQ and it is evaluated by the team comprising of members from the D&E Directorate. At least three major counterinsurgency training facilities, as they are known in the Army, were set up by Kiyani at Jharri Kas (near Kharian), Mangla and in Pabbi Hills and are used for the PIT (Zaki, 2019,

Interview, 20 January). Troops are introduced to the terrain and operations in FATA/ Balochistan while undergoing training at these facilities (Zaki 2019, Interview, 20 January). Mockup villages and replicas of streets and housing have also been set up at the training sites to train troops in clearing operations. PIT is still in vogue in the Pakistani Army (Qadeer 2022, Interview, 08 December). As a matter of policy by GHQ, it is mandatory for each unit to undergo PIT and pass the evaluation tests before induction into the counterinsurgency operations. Zaki highlights that "the IGT&E Branch, GHQ keeps on updating, from time to time, the curricula of PIT cycles as per the changing operational requirements of the sub-conventional operations (SCO). Same goes for the corresponding evaluation system of PIT" (2019, Interview, 20 January).

The initiatives regarding COIN teaching, training, and practical manifestation of the SCW doctrine (2013) were carried forward by the successors[1] of Kiyani, especially General Qamar Bajwa. Bajwa gave it a new outlook at the operational level in 2016 by resorting to the strategy of Intelligence Based Operations (IBOs) in Balochistan (Kamran 2018, Interview, 15 December; Jaffer 2019, Interview, 17 January). These were based on sophistication in accurate, timely and actionable intelligence. Interagency coordination for appropriate, effect-based, and focused use of force was set to be the hallmark of the IBOs. This strategy, in line with the SCW doctrine (2013), resulted in very positive outcomes such as a significant drop in casualties during the combat operations in Balochistan, etc. IBOs will be thoroughly discussed in Chapter 6. That is how the new COIN culture of the Pakistani Army was developed, as Edgar Schein (2004) maintains "one way that leaders transmit and embed culture is through what they teach and what they put forward as best practice, this could be both practical teaching [training], and doctrinal or academic teachings, which represents the institutionalisation of military culture" (cited in Nolan, 2012, pp. 97–98). Figure 4.1 (below) summarises this learning cycle.

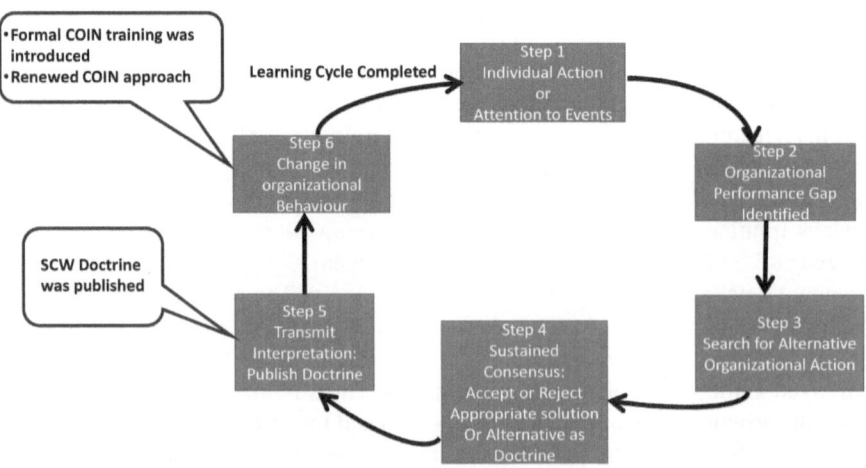

Figure 4.1 Pakistani Army's Learning Cycle (September 11, 2001–2013).

Source: Author.

**Military Organisation Learning, Innovation, and Adaptation
of the Pakistani Army in Relation to Counterinsurgency**

Military Innovation and Adaptation

The scholarly discourse on military innovation studies has advanced substantially since 1984 when Barry Posen published his seminal work *The Sources of Military Doctrine*. Since then, *how militaries innovate* has been an important theme of the military and strategic studies literature. This literature has increasingly focused on the driving factors of military innovation, that is, how militaries evolve new concepts/doctrine, force structures, plans, operations, and other aspects of the military metier to improve military performance while facing operational challenges. Military innovation is often identified as "a change in operational praxis that produces a significant increase in military effectiveness" (Barnett, 1963, p. 11). The driving factors that influence the course of military adaptation and innovation include changes in the external environment, civil-military relations, the role of leadership (both civilian and military), the sway of well-placed individual "mavericks" within the organisation, organisational culture, and alliance and domestic politics (Ucko, 2009, p. 16). The focus here is: "What exactly constitutes a military innovation"? (Grissom, 2006, p. 906). According to Grissom, three tacit assumptions are highlighted by the literature on innovation in military studies. Firstly, innovation changes the functioning of military field formations. Secondly, innovation is significant in scope and impact. Thirdly, innovation is associated with significant military effectiveness (Grissom, 2006, p. 907). Grissom argues that owing to these tacit assumptions researchers mostly focus on top-down approaches in order to explain military innovations by using the following lenses: (a) civil-military relations; (b) interservice politics; (c) intraservice politics; and (d) organisational culture (2006, p. 908). These are also termed as the four primary schools of thought of the military innovation field[2].

**The Civil-Military Model of Military Innovation/Learning and the
Case of Pakistani COIN**

Posen is considered to be the primary architect of the "civil-military model of military innovation" which he codified in his book *The Sources of Military Doctrine*. He studied the empirical cases of the doctrinal development of Britain, France, and Germany during the interwar years. Posen (1984, pp. 232–235) reached the conclusion that the innovation of militaries is determined essentially by the civil-military dynamics. He further highlighted as a part of his conclusions that intervention in doctrinal development, ideally through "maverick officers"[3], within the service will lead to innovation (Posen, 1984, pp. 222–236). Many empirical studies buttressed Posen's ideas of the civil-military model of military innovation (Grissom, 2006, p. 909); for example, Avant (1993; 1994) and Zisk (1993, p. 183) also find support for the civil-military model of innovation.

Civil-military relations in Pakistan have a chequered history owing to the fact that the country had been under military rule, where the COAS were also the presidents, for over 30 years (Hashim, 2013). Huntington's theorised "objective civilian control" equilibrium, which underscores the military's "functional imperative" of providing national security without subverting the political (democratic) system of government (1957, p. 83, 94), could not be maintained in Pakistan. The objective control theory implied a questionable claim of "moral neutrality" and did not consider the "power aspirations among the military" (Abrahamsson, 1972, p. 149, 163) as historically had been the case with the Pakistani Army. In 1958, General Ayub Khan seized power from the civilian government in Pakistan through a coup d'état and imposed the first-ever Martial Law in the country (Finer, 2002, p. 2). From this point onwards, the country was under intermittent military rule. Moreover, Pakistan is a country with "low political culture" so public awareness about political institutions is small (Finer, 2002, p. xvii). This is also one of the reasons that its military had ruled Pakistan for so long.

Coincidently all the rounds of the Balochistan insurgency, except the first round in 1948, were either initiated or fought mainly during the military rule in the country. Moreover, the civilians, irrespective of the executives or politicians (parliamentarians), were kept at bay while planning and executing the counter-insurgency during the military rule and this stands true for all the insurgencies that the Pakistani military fought during the dictatorial military regimes as earlier highlighted in the book. Hence there was no chance of military innovation concerning counterinsurgency through civil-military dynamics. Instead, the lack of civil-military synergy has been amply figured out in the book as a weak link in the institutional learning of the Pakistani Army. This holds true for other insurgencies, fought by the Pakistani military, as well such as the East Pakistan Insurgency in 1971 and FATA insurgency from 2001 to August 2008.

The civil-military synergy which contributed towards the learning and finally promulgation of the SCW doctrine of the Pakistani Army started to gather momentum in mid-August 2008 after the return to civilian politics, once General Musharraf stepped down as the COAS and later as President. The political ownership of the counterinsurgency efforts in FATA was clear from Operation Rah-e-Rast (*The Correct Path*) in Malakand and Swat in 2009 onwards. General Kiyani appeared as the "military maverick" of that time with regards to the innovation and learning of the Pakistani Army. Although the civilian leadership did not dictate to Kiyani exactly what doctrinal developments they wanted, he was asked by the then civilian political leadership to devise ways and means, organisational changes and a wholesale strategy to eradicate the ongoing insurgency in the country (Muzammil, 2018, Interview, 27 December; Sardar Nasir 2019, Interview, 10 January). Posen (1984, p. 54) argues that political leaders must compel military leaders as they are reluctant to change, to carry out and implement their desired changes. In this case, General Kiyani was not reluctant for the change in the counterinsurgency approach/ praxis of the Pakistani Army. General Kiyani and his team were already focused on conceptualising the intellectual aspects of Pakistani COIN as "how the next war [future counterinsurgency campaigns] will look like and how the army should prepare to fight it". This situation aligned well with Rosen's postulate (1991, p. 20) that innovation processes commence when the senior officers begin pondering the explanation of future war and the way to fight these in order to win. The go-ahead by the civilian leadership gave an impetus to the already existing idea in the minds of the senior command of the Pakistani Army of that time.

As Posen (1984, p. 54) posits that the mavericks are supported and promoted by the civilian leadership, so was the case with Kiyani. General Kiyani was granted a further three years' extension of tenure as COAS in July 2010 (MacDonald, 2010) by the then government after getting endorsement form the cabinet. The extension of tenure of the COAS by the democratic government in Pakistan was an unprecedented move. Moreover, the military defence budget was enhanced by the Federal government by about $1.28bn (£810 million)[4] (BBC News, 2010). This aided greatly in restructuring, revamping, and uplifting various directorates within General Headquarters (GHQ) Rawalpindi, such as Doctrine and Evaluation (D&E) Directorate, Military Operations (MO) Directorate, Inspector General Training and Evaluation (IGT&E) Branch and Inter Services Public Relations (ISPR) Directorate as highlighted earlier in the chapter. These directorates played a pivotal role in conceptualising and promulgating the COIN doctrine, designing, and implementing COIN training regimes and formulating the evaluation mechanisms for the COIN training as highlighted earlier in the chapter. Moreover, the regular budgetary increments also helped in establishing various training facilities for the COIN operations as already highlighted. All these steps greatly assisted in the promulgation of the COIN doctrine of the Pakistani Army reflecting a renewed COIN approach and later its practical manifestation through extensive education and training as already argued in the chapter. Posen (1984, p. 14) considers military doctrine as a vital indicator of military innovation. That is how, unlike the earlier three learning cycles, the fourth learning cycle of the Pakistani Army could be completed. Hence the following assertion of Zisk holds true in the case of the Pakistani Army as well "… if a military is to implement change, it must acquire political support" (1993, pp. 4–5).

The Cultural Model of Military Innovation/Learning and the Case of Pakistani COIN

The cultural model of military innovation is another school of thought in military innovation studies. However, it has not yet attained the same interest levels and stature as the other more established schools in the field (Grissom, 2006, p. 916). Theo Farrell is the key scholar who contributed to this aspect. Farrell (1996; 1998) and Farrell and Terriff (2002, pp. 7–8) put forward the idea that organisational culture attains the status of a main causal factor in military innovation. Farrell defined culture as "intersubjective beliefs about the social and natural world that define actors, their situations, and the possibilities of action" (2002, p. 49). Farrell and Terriff argue that military innovation is one of the pathways of military change whereas they defined the latter as "change in goals, actual strategies, and/ or structure of a military organisation" (2002, p. 5). They further highlight that the analysts also need to consider processes within a military organisation to explain military change (Farrell & Terriff, 2002, p. 16). Farrell's research sums up that the three ways culture can impact military innovation: a) senior military leaders can lead the organisation towards innovation drive by transforming the culture; b) external shocks can impact the culture and thus provide an enabling environment for innovation; and c) by imitating other militaries which he termed "emulation"[5]. Kier (1995; 1997, pp. 72–77) has also pushed forward the concept of culture as an explanatory variable of doctrinal innovation.

By following Farrell and Terriff's suggestion of considering processes within a military organisation to explain change and consequently analysing various institutional learning cycles of the Pakistani Army, one major factor stands out: the role senior military leadership in reshaping culture to lead to COIN innovation or its lack thereof. For example, the third learning cycle of the Pakistani Army could not be completed because, apart from other reasons as highlighted in Chapter 3, the senior military leadership, like General Zia ul Haq (COAS and the president of Pakistan) and General Akhtar Abdur Rahman (Chairman Joint Chiefs of Staff Committee (CJCSC) of the Pakistani Armed Forces) and their successors, were traditionalists and believed that conventional warfare was the main role of the Pakistani Army. Therefore, they did not reshape the culture of the Pakistani Army to lead the organisation towards COIN learning and innovation. Conversely, General Kiyani and his successors, mainly General Raheel Sharif and General Qamar Bajwa, along with their team endeavoured to reshape the culture of the Pakistani Army with regards to counterinsurgency as highlighted in the current chapter and will be further dwelled upon later in Chapter 6. This resulted in the promulgation of the SCW doctrine and later on its practical manifestation on the ground (which will be thoroughly covered in Chapter 6). A new emphasis on the SCW doctrine also played a significant role in redefining the officers and troops of the Pakistani Army. Deep cultural changes were embedded in the rank and file of the army via professional military education and extensive training in counterinsurgency. As Mullins asserts, the curriculum of the military educational/training institutions will play a significant role in shaping the professional military culture and "therefore influence the direction of military innovation" (2000, cited in Grissom, 2006, p. 919). Victoria Nolan sums up the debate about top-down military innovation models very succinctly: "There is an emphasis on both culture and leadership [both military and civilian] as the critical enablers of learning and innovation …. and [hence] imply a connection between culture and leadership" (2012, p. 20).

Military Adaptation and the Case of Pakistani COIN

Grissom (2006, p. 930) and Cohen (2004), while analysing top-down approaches, explicitly underscore that there is a dearth of research on bottom-up approaches to military innovation. Grissom suggests "there is an entire class of bottom-up innovations that have yet to be explored, understood, and explained" (2006, p. 930). Farrell (2010, p. 569) has precisely differentiated adaptation from innovation where the former is defined as bottom-up "change to tactics, techniques, or existing technologies to improve operational performance" and the latter is defined as top-down "major change that is institutionalised in new doctrine, a new

organisational structure and/or a new technology". After operationalising the concept of bottom-up adaptation, Farrell argues that adaptation may involve "exploitation of existing tactics, techniques and technologies, or it may involve exploring
new ways and means of operating" (2010, p. 590).

In the Pakistani Army, organisational learning also occurs through adaptation by
the units/formations at a tactical and sub-tactical level even though adaptation is not
a sufficient condition for it. Usually learning through adaptation stays for a short
time and has limited impact. In most of the cases, organisation-wide learning owing
to adaptation is not possible as individuals, and clusters of units may adapt and learn
but once personnel leave the subject unit/formation, the unit may forget the insights
gathered through learning (Catignani, 2014, p. 38). Thus, the operational improvements carried out as a result of localised adaptation are usually lost (Farrell, 2020,
p. 6). This "organisational forgetting" is troublesome as the units again revert to old
habits. The personnel of incoming units to the operational areas "relearn lessons
learned" and "reinvent the wheel" during their deployment[6]. This has not been the
case with the Pakistani Army. The informal COIN learning in the form of lessons
learned and improvements obtained in the localised context through adaptation is
preserved through the existing mechanisms in vogue. War Diary and the Formation
Training Note are two essential documents which help in preserving the knowledge
gained through adaptation in the knowledge repositories of the Pakistani Army.
These two documents are maintained and updated very religiously throughout the
army, again a British Army's legacy of the colonial era. The war diary has a section on operations and is maintained even at a company/squadron level in each unit
and formation. The new knowledge gained or lessons identified through localised
adaptation (even minute details) are recorded in the war diary and endorsed by the
respective company and squadron commander.

The operational contents of these war diaries are sent directly to the D&E
Directorate and the local formation on a quarterly basis as a part of routine reports and returns. The contents which are valid or useful for the entire organisation are circulated throughout the army in the form of Army Operational/Training
Instructions or included in the operational/training manuals but only after validation by the Research and Development (R&D) Directorate. The operational
contents, which are geographic area/formation specific, are included in the formation training note after validation by the operational staff of the concerned formation. The formation training note is the key document which guides the training of
the incoming/newly inducted units into the formation's operational area. It acts as
a refresher in training cadres of the units on the formation's order of battle and in
designing the curriculum of the respective Formation Battle Schools. That's how
the knowledge acquired through adaptation is preserved in the repositories, passed
on and practised by the newly inducted unit of the concerned formation. For example, Brigadier Akber Khan, the commander of the forces in Kalat in 1948 during the first round of Balochistan Insurgency, endorsed his observations in the
war diary of the Brigade HQ regarding the requirement of a clear interpretation
of concepts of the "use of minimum force" and "use of necessary force" and the
"rules of engagement". Likewise, Lieutenant Colonel Mitha in the 1960s studied

the operations against Ipi Faqir in Waziristan in (1952–53 and 1954) through the unit war diaries. This helped him in getting insight into these operations of Frontier Warfare which were conducted for the very first time by the Pakistani Army/FC after independence. Moreover, while developing the SCW doctrine for the Army, the D&E Directorate thoroughly consulted the lesson learned notes and war diaries of all the units that remained actively involved in internal security operations, operations in aid to the civil power and the counterinsurgency operations since 1947. This included the war diaries of the Frontier Force (FF) units which participated in various campaigns in FATA/NWFP before and after the partition of India. The D&E Directorate took considerable time in carrying out this massive task.

The COIN learning process of the Pakistani Army comprised of both top-down innovations and bottom-up adaptation activities while the former dominated this learning. The adaptation activities were codified into SOPs, formation training notes, army operational/training instructions and operational/training manuals, and hence became innovations and subsequently altered the way in which the Pakistani Army fought the insurgency in Balochistan during the fifth round after the promulgation of the COIN doctrine. The organisational learning of COIN in the Pakistani Army was institutionalised through publications (mainly doctrine), changes to existing curricula, developing new training regimes, designing the evaluation mechanism of COIN, training and most importantly the delivery of formal learning programmes – mainly extensive training and thorough education of both the officers and the troops.

SCW Doctrine

SCW is a generic term encompassing all armed conflicts, below the threshold of a conventional war between states, like militancy, insurgency, proxy war, and terrorism employed as a means in an insurrectionist movement or undertaken independently (Dixit, 2010). Such conflicts are political wars in the most literal sense, so they require the synergy of efforts between the armed forces and the other instruments of national power. Hence any meaningful effort to resolve or manage the conflict is invariably long term and involves a multi-agency civil-military response backed by socio-political and politico economic manoeuvres.

The doctrine (2013) defines SCW as a conflict between the violent non-state or sub-state actors and the instruments of the state, involving the use or threat of the use of force. Quite often, the sub-conventional threat, after the state has failed to contain it through its civil law enforcement machinery, has the proclivity to transcend to a level wherein the involvement of armed forces is warranted. The kinetic or non-kinetic employment of armed forces in SCW environment is referred to as sub-conventional operations (SCO) that are aimed at regaining a state of "normalcy" so that the civil authority can re-assert itself in the post-conflict stage (SCW doctrine, 2013, p. 1).

The SCW doctrine (2013) is the synthesis of the Pakistani Army's ethos of fighting *small wars*, inherited from the rich experience of the British Indian Army. However, it developed its own diverse COIN experience while fighting in Balochistan, East Pakistan and FATA/NWFP over a period of six decades. In other words, it has emerged from the *small wars*/counterinsurgency experience, moulded and innovated as per the modern requirements of the Pakistani Army through the systematic analytical process/learning cycle as already described in detail in the last chapter. Thus, the SCW doctrine (2013) sets the counterinsurgency approach of the modern-day Pakistani Army, which is the reflection of the organisational culture. Schein highlights an important point, regarding doctrine as a rendition of organisational culture, "at the most visible level, an organisation's documentation epitomises the very tangible overt manifestation [of culture] that one can see and feel" (2004, cited in Nolan, 2012, p. 98). Nolan further observes that "in the military, doctrine is one of these observable cultural artefacts, which articulates the army's beliefs and values, along with its ethos and guiding principles .... [in this case for counterinsurgency in the form of SCW doctrine]" (2012, p. 98).

The SCW doctrine (2013) was the second doctrinal document the Pakistani Army produced. In November 2011, the Pakistani Army promulgated the *Pakistan Army Doctrine* (PAD) for the first time in the history of the army. PAD (2011), again an initiative of General Kiyani, was meant to provide high-level doctrinal guidelines to all components of the army and a framework for the development of the army for comprehensive planning and operating successfully in the new spectrum of conflict. PAD (2011) highlights the strategic environment, defines the doctrinal facets of fighting a conventional war and asymmetric operations, keeping in view the modern technology and revolution in military affairs. Moreover, it presents the principles of warfighting. PAD (2011) recognises that the basic role of the Armed Forces is the defence of the country from external threats: that is conventional warfare. However, the military capability developed to deliver this primary role is frequently tasked to perform a variety of subsidiary and auxiliary roles. As per PAD (2011), depending on the environment, auxiliary roles such as sub-conventional operations (SCO) often take pre-eminence and as such their priority or significance remains dynamic. That's how PAD (2011) addresses US concerns, as mentioned earlier, about the counterinsurgency operations allegedly being a low priority for the Pakistani Army in the backdrop of its primary focus on fighting a war against India. PAD (2011, p. 5) segregates the employment of the army into two distinct categories a) Defence Regime and b) Security Regime. The former covers all aspects related to the defence of the country from external threats and involves conventional and strategic forces whereas the latter embodies a wide spectrum of operations other than war (mainly SCO which includes kinetic actions against the insurgents, Aid to Civil Power, and internal security operations). The Security Regime Operations, whether kinetic or non-kinetic, have the common characteristic of the military providing assistance to other government or non-government departments and agencies. The SCW doctrine (2013) flows from the PAD (2011) and mainly focuses on the principles and best-suited practices, developed over a very long period in the context of the Pakistani Army, for successfully conducting the SCO.

The SCW doctrine (2013) provides guidance in three broad areas, namely security and military operations, political, and legal aspects and finally the civil-military coordination (CIMIC) of the sub-conventional warfighting. This corresponds to the three pillars of Kilcullen's Counterinsurgency Model. The SCW doctrine (2013) also highlights the importance of the Information Operations (IOs), which represents the base of Kilcullen's model. Apart from this, initially, the SCW doctrinal document (2013) conceptualises the spectrum of SCW and elaborates the existing environment and challenges for the Pakistani Army. This can be equated to Kilcullen's concept of the *Conflict Environment*.

The rest of the chapter will highlight the nature and conceptual contours of SCO as per the SCW doctrine (2013) while elaborating the doctrinal approaches to security and military operations, political and legal aspects, CIMIC, and IOs. During the discussion, the lessons learned/performance gaps identified in both the completed and broken learning cycles of the Pakistani Army since 1947, as described earlier, will be frequently referenced to demonstrate the evolution of the final counterinsurgency approach in the form of SCW doctrine (2013) or, in other words, the result of the learning process.

The Nature of the Counterinsurgency Problem

Conceptualisation followed by the promulgation of the formal doctrine for countering insurgencies depends on precisely establishing the nature of the problem. This process of determining the nature of the problem took a long time to evolve in Pakistan as described in Chapter 3. Starting from the initial Operations in Aid to Civil Power in Kalat (1948) and counterinsurgency campaigns against the Faqir of Ipi in FATA (1952) until mid-August 2008, the Pakistani Army generally adopted the strategy of "butcher and bolt" inherited from the British Indian Army. The Pakistani Army followed the enemy-centric approach largely focused on annihilating the enemy and its support while relegating the political and moral dimensions to the overall COIN strategy (Asim 2019, Interview, 25 January; Adeel 2019, Interview, 08 January). Through hard experiences, the quest for improving the outcome of COIN campaigns, and as a result of General Kiyani's initiatives, the Pakistani Army accepted Kitson's idea that "there can be no such thing as a purely military solution because insurgency is not primarily a military activity" (Kitson, 1977, p. 283). They accepted that counterinsurgency has to be fought simultaneously on multiple fronts including the political, economic, military, psychological, and social while the political aspect takes precedence (Adnan 2019, Interview, 14 January; Fahad 2022, Interview, 29 January) as the "State's response to various forms of sub-conventional threats primarily resides in the political domain" (SCW doctrine, 2013, p. 22). The adoption of Kitson's position by the Pakistani Army dictates that they acknowledge the fact that military means is only one of the elements, possibly at times the least important one, of the comprehensive response to the counterinsurgency.

Instead of military victory on the battlefield, the Pakistani Army now accepts that the SCO is aimed at deterring, dissuading, or defeating organised defiance to law and authority and is primarily designed for reconciliation, reintegration, and rehabilitation (SCW doctrine, 2013, p. 9; Nadeem 2018, Interview, 13 December). It aims to manage violence through political initiatives such as dialogue, negotiations, Disarmament, Demobilisation, and Reintegration (DDR) programs and reconciliation opportunities for the militants. In addition, for the sustenance and success of the political initiatives, any military action has to be short-lived, effect-based with the least collateral damage (SCW doctrine, 2013, p. 9). The COIN operations should have well thought out political objectives; otherwise, they will not yield any desirable results like sustainable peace or a permanent end to violence. The political and military objectives can be achieved by addressing the strategic centre of gravity, that is, the population in this case.

The SCW doctrine (2013) adopts the population-centric approach for the SCO of the Pakistani Army. It highlights the fact that more often in SCW, the population emerges as the centre of gravity. Instead of incapacitating the population, it has to be protected, reinforced, and empowered to function closely with LEAs (SCW doctrine, 2013, p. 24). The population must be convinced that the operations against the insurgents are for their safety and security. The operations must be developed keeping in view the local conditions, environment and most importantly the aspirations of the population. Maintaining or recovering the support of the population should be kept in focus through all the stages of operations (SCW doctrine, 2013, p. 24). The SCW doctrine (2013) clearly addresses the apparent lack of population centricity and the use of heavy kinetic means as the sole option as exhibited by the Pakistani Army during the East Pakistan campaign, earlier rounds of Balochistan insurgency and the initial operations in FATA as well.

SCW is not primarily about killing or capturing insurgents, but about winning hearts and minds and changing the perception of the population towards insurgency. The failure of the government to ensure the security of the local population and address their grievances will allow militants to exploit the situation to their advantage. If that happens and the population turns to support the militants, the population could extend all kind of human and material support. The population, therefore, emerges as a core element, support of which to either side (militants or government) is vital for determining success. This calls for a very thorough and well-planned inter-agency counterinsurgency operation focusing with simultaneity on the security, political and economic aspects along with comprehensive IOs. The Pakistani Army now believes that collateral damage is always counterproductive as it leads to a rise in insurgency and reduces the space for the political initiatives which are necessary for managing and resolving the conflict (Ashfaq, 2016; Qadeer 2022, Interview, 08 December).

Addressing the Cause of the Insurgency and the Conflict Environment in Balochistan

All major insurgencies are driven by some strong cause(s) which motivates the people to fight against the state (Banerjee, 2009, p. 198). In the case of Pakistan, all the past insurgencies were caused as a consequence of marginalising particular ethnic communities resulting in socio-political disparities, a fragile economic situation and a lack of development. The one exception to this was the insurgency in FATA as a result of the GWOT. The principal motive behind the FATA insurgency was the global jihadist trend, and it surfaced as retaliation against Pakistan's decision to join the coalition in the GWOT. These causes of insurgencies are easily understandable as the political victimisation of a particular community, and lack of economic development in a specific area gives rise to sub-nationalism, which is often referred to as pseudo sub-nationalism by the Pakistani Army (Asim 2019, Interview, 25 January; Kazmi, 2018, Interview, 19 December). The insurgents exploit these causes to gain popular support and sometimes get external help. In such a scenario, people opt for armed resistance for independence so that they can control their political and economic fate. A case in point is the successful East Pakistan insurgency where Indian assistance in training and equipping the Mukti Bahini played a significant role in the Bengalis gaining independence. Moreover, during all the rounds of the Balochistan insurgency (except the Kalat uprising in 1948 against the forceful merger of the Independent State of Kalat to the Federation of Pakistan) the principal causes were political victimisation and a lack of economic development in the province.

In such a "conflict ecosystem" linked to an insurgency, many actors prevail, each with a specific set of objectives (at times with the help of external support). These actors compete with each other. As Kilcullen puts it:

> .... each [actor] seeking to maximize their own survivability and advantage in a chaotic, combative environment .... these actors are constantly evolving and adapting. They include government, ethnic, tribal, clan or community groups, social classes, urban and rural populations, and economic and political institutions.
>
> (2006, p. 2)

The Pakistani Army's doctrine highlights SCW as a "Clash of Wills" between two primary actors, i.e., militants and the government, whereas population support emerges as a core element of success for both. "External Support" is usually crucial for the sustenance and success of the resistance (SCW doctrine, 2013, p. 7). A clear understanding of this framework is essential for achieving the objectives of SCW.

The security milieu of Pakistan has always remained fragile with the recurrent emergence of challenges in the sub-conventional domain; compounded by competing interests of outside powers and their nexus with internal actors (Adnan 2019, Interview, 14 January; SCW doctrine, 2013). The examples from the Pakistani COIN experiences, as discussed above, demonstrate why the SCW doctrine highlights, "political factors which nurture SCW ... a) political instability b) ineffective state institutions c) weak governance and d) sub-nationalism" (2013, p.12). While elaborating on the sub-nationalism, the doctrine stresses that the political disharmony and unequal distribution of resources have given rise to the sense of deprivation amongst smaller provinces (referring mainly to Balochistan and KPK (including FATA)) (SCW doctrine, 2013, p.12).

The situation in Balochistan is a case in point where the failure to administer and afford fair developmental opportunities has created serious dissatisfaction amongst the people. There is an awareness amongst junior officers in the army that the military cannot address the root causes of insurgency in Balochistan through kinetic means alone; its job is to facilitate the political process and economic growth by controlling the violence by insurgents and assist the civil authorities in gearing up the development process in the province (Kamran 2018, Interview, 15 December). They believe that the successful implementation of CPEC in Balochistan if it sincerely safeguards the economic interests of the province, will help a great deal in ruling out one of the root causes of insurgency in the province.

Use of Force and Role of the Military

Appropriate, Effect Based, and Focused Use of Force to Avoid Collateral Damage

The Pakistani Army has a long history of resorting to excessive use of force while conducting Operations in Aid of the Civil Power and Counterinsurgency. It can be attributed to the legacy of the ethos of the British Indian Army from whom the Pakistani Army traces its origin as discussed in Chapter 3. The former resorted to excessive use of force sometimes in the name of "necessary force" and other times the "provision of adequate force" appeared as the fundamental principles in the operational manual (*Duties in Aid of the Civil Power*, 1937). Adherence to these principles was demonstrated by the British Indian Army during Gandhi's Quit India Movement (1942–44). The dilemma of striking a balance between the excessive and minimum use of the force in independent Pakistan was first highlighted by Brigadier Akber Khan during the Kalat Operations in 1948 as discussed earlier. In the early 1960s, Mitha highlighted that collateral damage was one of the outcomes of the excessive use of force. He argued that the Pakistani Army should understand that the insurgents were citizens of Pakistan. He proposed the concept of an "effect-based approach" to the use of force. However, this was not adopted by the Pakistani army, as discussed in Chapter 3, as the learning cycle was broken by the conventional war with India in 1965. However, the dimension of "Appropriate, Effect Based and Focused Use of Force to Avoid Collateral Damage" in the SCW doctrine (2013, p. 24) reflects that learning has now taken place in the light of Army's experiences of fighting insurgencies.

The COIN principle of "appropriate, effect-based and focused use of force to avoid collateral damage" is fundamental in the context of deploying the military against the insurgents who are Pakistani citizens as well. The SCW doctrine maintains that "the force to be used against the militants must be appropriately scaled, effect-based and focused in the application" (2013, p. 13). It is one of the objectives, along with providing a reasonable level of security for other state organs to re-establish, reassert and operate for the restoration of normalcy, set in the doctrine for the Pakistani Military engaged in SCOs. The doctrine asserts that disproportionate and wanton use of force may undermine public support at any stage of the operation. Therefore, the military commander must never use more force than is absolutely necessary to achieve the objective. This may also entail minimum collateral damage. As part of the specific guidelines on the conduct of SCO, the SCW doctrine (2013) denounces collateral damage. Therefore, it addresses the problems regarding the balance between the excessive and minimum use of force and the collateral damage earlier highlighted by Akber and Mitha.

Comprehensive Strategy at National Level

Earlier the employment of the Pakistani Military in counterinsurgency was devoid of any comprehensive strategy at the national level (Kazmi 2018, Interview, 19 December). This holds truer for the operations conducted during the dictatorial military regimes like East Pakistan insurgency in 1971 during the reign of General Yahya Khan as the president and more recently in 2006 in Balochistan during the rule of General Musharraf. During these operations, the politicians were marginalised, and there was no comprehensive national strategy formulated except the Military Strategy at GHQ, Rawalpindi. Though some economic reforms were introduced during the latter insurgency, those were again implemented through the military establishment in Balochistan (mainly through Pakistani Army's HQ Southern Command, Quetta) without taking the Provincial Government into confidence. Thus, these economic reforms were unable to win the hearts and minds of the people. Another important aspect of the SCW doctrine (2013, p. 23) is that it elaborates guidance on the conduct of military operations with the premise that the application of the military instrument takes place in the broader framework of the National Security Strategy which comprises the political and economic dimensions as well. The doctrine highlights the importance of creating military operations acceptable to the public at large with political ownership, winning public confidence, encouraging local participation, and finally creating a friendly environment for sustained development and bringing about better socio-economic change as the preconditions to be kept in mind for employing security forces in SCW.

Spectrum of Military/Security Operations

SCW will involve a variety of military actions. Some of these operations will be referred to in the next chapters of the book, so it is pertinent to explain the relevant operations as per the doctrine. These operations are also widely carried out by the Pakistani Army in Balochistan as a result of renewed COIN approach.

Broad Spectrum Security Operations (BSSOs)

BSSOs involve multiple lines of operations such as a sting or snap operations, combing operations, patrolling, route piqueting, a show of force and establishing joint check posts etc. by the Army itself or by other LEAs under the security umbrella provided by the armed forces (Kazmi 2018, Interview, 19 December). Most of these operations are the intelligence based operations commonly abbreviated and known as IBOs in the Pakistani Army and media (Jaffer 2019, Interview, 17 January).

The Pakistani Army generally suffered from a lack of timely and actionable intelligence throughout its history of counterinsurgency operations. That is why it could not successfully plan and execute IBOs in the counterinsurgency environment. The SCW doctrine (2013) lays very strong foundations by highlighting the importance of intelligence. It maintains that SCOs are mainly intelligence driven BSSO and continue through all stages of the conflict including combat operations (SCW doctrine, 2013, p. 31). Thus, timely, accurate, and actionable intelligence, through an integrated approach by all civil and military intelligence agencies is the key to success. The focus of intelligence efforts should be to ascertain maximum information about the infrastructure of the militants, their ideology, domestic, and foreign support and support base in the civil population. All such kinds of intelligence based BSSO contribute towards the overall success of the campaign by achieving some or all of the effects, like sapping militants' will, curtailing militants' liberty of action, enhancing public confidence in the military's capability that leads to more support base within the population in favour of the military and preventing collateral damage (SCW doctrine, 2013, p. 34).

The IBOs in the recent past by the Pakistani Military proved to be very successful and achieved most of the above-narrated effects. A case in point is "Operation Radd-ul-Fasaad" launched by the Pakistani Military in coordination with Civil Armed Forces (CAF) and other security and law enforcing agencies (LEAs) in February 2017 across the country including Balochistan (Dawn, 2017). An ISPR (2017) press release No PR-87/2017 described "Operation Radd-ul-Fasaad" as the Broad Spectrum Security Operation for eliminating the residual threat of terrorism; consolidating the gains made thus far by military operations under Operation Zarb-e-Azb; and de-weaponisation of the society. In the first six months after the launch of "Operation Radd-ul-Fasaad", around 9000 intelligence based operations were carried out across the country which led the army to launch 46 major operations against terrorists across the country (Siddiqui & Akbar, 2017). The decreasing numbers of civilian causalities and declining trend of terrorist incidents in the country during this operation were the indicators of the success of "Operation Radd-ul-Fasaad". (Siddiqui & Akbar, 2017; Ullah, 2017).

Information Operations and Media

Effective Information and Psychological Operations and effective media handling were great voids, owing to the absence of a formal policy on the subject, throughout the history of the Pakistani Army's counterinsurgency campaigns. The SCW doctrine (2013) lays out detailed guidance on the subject by establishing that the Information Operations (IOs) have a predominant role throughout the sub-conventional conflict continuum. IOs focus on all segments of society involved directly or indirectly (the neutral majority, pro or anti-state elements as well as the militants) with well-conceived narratives and counter-narratives using different means and mediums. While the broader themes and narrative are orchestrated at the state level, the IOs conducted by the military during SCOs must be in sync with the state narrative. IOs consist of four core components a) military security and deception b) Psychological operations c) electronic warfare d) support to public diplomacy (usually an extension of Civil-Military Cooperation) (SCW doctrine, 2013). The doctrine categorically highlights that dominating the information domain is not solely the purview of the military alone but all the elements of national power. In the IOs the spirit of political dominance is also maintained as the SCW doctrine states: "The political authority should frame an overarching and well synchronised national strategy, primarily through IOs, for shaping the environment in their favour" (2013, p. 36). Essential steps for shaping the environment may include a well-articulated IOs campaign to legitimise the application of the military's response, winning public support, and confidence through perception management and effective media handling to achieve the moral high ground. Perception management of the local populace in particular and the national and international community in general stand out as an important facet of IOs.

The Pakistani Army committed many blunders in the past owing to the absence of a media policy at the national and international level and countering the foreign media (Ahsan 2019, Interview, 16 January) e.g., the expulsion of foreign journalists from East Pakistan in 1971. During operations in FATA (2002–2010) the officers of ISPR had no formal training for utilising the media as a tool for the Psychological Warfare (Nawaz, 2011). To address this performance gap, the ISPR Directorate was revamped and reorganised. It now has a broader and more active role for perception management. Its budget was enhanced. It was upgraded from a side-lined military establishment with the group of officers including the DG with no prospects of future promotion (on the verge of retirement) to an active military establishment with an exuberant set of officers having a bright military career ahead (Nadeem 2018, Interview, 13 December; Muzammil, 2018, Interview, 27 December).

The SCW doctrine (2013, p. 46) also vividly highlights that the Inter Service Public Relations Directorate (ISPR) in coordination with MO Directorate must work in close harmony with the national media to support SCOs. It recognises that the media plays a vital role in perception management. Media can be used for countering the militant ideology, garnering public support for operations, highlighting the illegitimacy of a militant's cause, ensuring effective coverage of the Army's achievements along with its sacrifices to build national will and morale and support post-operation rehabilitation and normalisation initiatives. The efficacy of IOs and the role of ISPR in designing the media strategy for Balochistan will be critically examined in Chapter 6 of the book.

Winning Hearts and Minds (WHAM)

Winning hearts and minds (WHAM) in the COIN campaigns is a tricky proposition as it involves a lot of coordination and synergy of efforts, along with requisite funding for the projects, between military and civil agencies in different areas of operations. Adnan (2019, Interview, 14 January) argues that the Pakistani Military realised the importance of WHAM during the operations in FATA. Prior to that, there was no significant example of a consolidated winning hearts and minds programme by the Pakistani Military in any insurgency ridden area except for few odd steps taken in isolation from the overall coercive strategy without any follow-ups or requisite framework for sustained development e.g., General Zia's amnesty to Baloch prisoners arrested by Zulfiqar Ali Bhutto's government on treason and armed violence charges during the fourth round of insurgency (1973–78) in Balochistan (Shaikh, 2014). This was a step for winning hearts and minds of the Baloch people, but unfortunately, General Zia could not introduce any sustainable development programme in Balochistan targeting the health, education or poverty alleviation as part of a well-thought-out WHAM initiative (Sardar Nasir 2019, Interview, 10 January; Jaffer 2019, Interview, 17 January).

With time the WHAM initiatives in terms of civic actions during and after COIN appeared as the basic precondition for success for the Pakistani Military. WHAM gathered considerable traction within the senior leadership of the Pakistani Army (Jaffer 2019, Interview, 17 January) that is why it emerged as one of the essential facets of the SCW doctrine. The SCW doctrine (2013) maintains that the WHAM campaign is carried out through all the stages of the SCO by employing various means such as the reconstruction, rehabilitation and restoration of essential services to the population living in the area of operations, etc., which may lead to the economic development of the area and people. As rehabilitation initiatives are long term initiatives, they have to be spearheaded by the political/civil administration. However, the army, due to its inherent organisational strength, can assist the civil administration in speeding up the reconstruction process to restore normalcy (SCW doctrine, 2013, p. 38). The Pakistani Army, with the help of the Provincial Government, local civil administration, and NGOs, has resorted to diverse WHAM initiatives in Balochistan including the opportunity for reconciliation to the militants (Khan Iftikhar, 2018) which is advocated by the doctrine and well in line with the population-centric strategy of the Pakistani Army. These WHAM initiatives will be thoroughly discussed in Chapter 6.

Unity of Command, Unity of Effort, and CIMIC

The Pakistani Army's COIN experiences highlight Unity of Command, Unity of Effort, and CIMIC as the sine qua non of fighting successful counterinsurgency campaigns. COIN campaigns are a complex grid of political (civil) and military actions which require synergy to deal with the insurgency problem effectively. Therefore, the doctrine (2013) implies that all the elements of national power dedicated to SCW should be employed cohesively while emphasising that politico-military components must have a convergence of effort. Thus, the unity of command and unity of effort through CIMIC is vital in dealing with COIN.

For the unity of command, the Pakistani Army follows the single commander model, as used in the British Malayan Insurgency, to whom all the civilian and military agencies employed in the specific area report. The commander is overall responsible for the campaign and has the necessary authority. The SCW doctrine categorically highlights "The overall command of the operations must rest with the army. However, it will be transferred to the civil administration after return to normalcy" (2013, p. 24). Unlike India and some other countries, there is no concept of a "Unified Command Approach" to tackle the counterinsurgency in the Pakistani Army. However, the Defence Committee of the Cabinet (DCC), headed by the PM and attended by the key federal ministers and the defence services chiefs, exists at the federal level. The DCC is the highest political body which steers national security policy and responsible for the national war effort.

The doctrine (2013) defines CIMIC as the coordination and cooperation, in support of the mission, between the military and civil actors, including civil authority, the national, or local population and NGOs. It encompasses collaboration and information exchange between military forces and the in-theatre civil actors. It also includes the essential dialogue and interaction between civilian and military actors to protect and promote humanitarian principles, avoid competition, minimise conflict, and when appropriate, pursue common goals. That is how it yields the unity of effort in the COIN campaigns. CIMIC works as a force multiplier because of its potential to release military resources for other operational tasks or by increasing the effectiveness and efficiency of ensuing military activity. It is also essential when switching from fighting to law-enforcement and development.

The absence of CIMIC was a consistent phenomenon observed throughout the history of the Pakistani Army's counterinsurgency campaigns, as thoroughly discussed earlier in the book. The SCW doctrine (2013) acknowledges it as the basic tenet during the planning and most importantly execution of SCO and generates very comprehensive guidelines on the subject.

Counterinsurgency – the Rule of Law and Ethical Considerations

COIN experiences and teachings often highlight the fact that the counterinsurgents should operate within the ambit of the law. Thompson (1966, pp. 50–58) proposes a set of five principles for countering the insurgency basing on his population-centric approach. The second and very important principle highlighted by him is that the government must follow the law (1966, p. 52).

The rule of law principle has attained added importance owing to the internal nature of all Pakistan's COIN campaigns. The Pakistani Military, while providing aid to the civil power, is governed by the constitutional obligation and the provisions of applicable statutory law. The provision is mentioned in the constitution of Pakistan (Chapter 2, article 245) and Army Regulations (rules 557 to 559). All these constitutional provisions and rules establish the *de jure* supremacy of the civilian/political authority. The army accepts this supremacy as the doctrine (2013, p. 32) states that the military means are applied only to create a conducive environment for the attainment of the political objectives, which is the restoration of normalcy through negotiated settlement/resolution of the conflict, and for political efforts to succeed.

The doctrine equally covers the legitimacy of the SCO. In the past, the legitimacy of the SCO operations had been one of the issues on which the army was severely criticised. For example, Puri (2012, p. 91) maintains that the legitimacy of the army operations in FATA as part of GWOT remained controversial. The SCW doctrine lays great stress on the adherence to the law, both national and international humanitarian law while conducting the SCO.

Along with the legal aspects, the SCW doctrine (2013) reflects upon the ethical aspects as well; something relevant and equally important to consider while operating in a sub-conventional environment. It emphasises that the application of the law must be studied in conjunction with Articles 4, 5, and 6 of the Constitution of Pakistan, which elaborates the aspects related to citizens' rights and obligations. It further elaborates that due cognisance must also be given to the application of force as enunciated in the Law of Armed Conflict, Code of Conduct for SCOs, and Customary International Humanitarian Law. Moreover, the Code of Conduct for SCOs and Customary International Humanitarian Law are attached as annexures to the doctrinal document for the guidance of the armed forces personnel.

Rules of Engagement (ROE)

The doctrine (2013) defines the Rule of Engagement (ROE) as the directives issued by a competent political or military authority that delineates the limitations and circumstances under which forces will initiate and continue to prosecute combat engagement with hostile elements being encountered. The absence of the new ROE for the Pakistani Army after independence was identified by Brigadier Akber Khan in 1948, during Kalat Operations, as something negatively affecting the performance of the army. This may also lead to the excessive use of firepower and resulting in collateral damage, the point highlighted by Mitha in the early 1960s as a reason for the performance gap. Later on, during operations in East Pakistan in 1971, Balochistan in 1973–78 and much later in operations in FATA/NWFP till 2008/09, the army followed the inherited culture of the ROE from the British Indian Army which was based on the heavy reliance on firepower. There was no central guideline around which the local commanders could frame the ROE. This resulted in a lack of uniformity and excessive reliance on firepower as late as 2012 (Ahsan 2019, Interview, 16 January).

The SCW doctrine (2013, p. 40) highlights the importance of the ROE as a *control mechanism* designed to prevent the inadvertent escalation of a situation and strive to follow the general precepts of the law. The pivot around which the new ROE revolve is that all possible efforts are made to avoid collateral damage. However, the security of troops should not be compromised. This entails right to use force in self-defence against attack or threats of attack by miscreants/militants but also the prohibition of unnecessary retaliation and ban on firing directly on the unarmed protestors.

Conclusion

The SCW doctrine reflects the renewed organisational culture. There was no SCW doctrine before 2013; the Pakistani Army fought the earlier counterinsurgencies basing on its ethos of fighting *small wars*, inherited from the rich experience of the British Indian Army. In 2008, after General Kiyani took over as the COAS in November 2007, the army realised the need for change in counterinsurgency approach as the existing coercive approach caused a lot of collateral damage. Moreover, it particularly turned the population of FATA/NWFP against the army and people from other areas of Pakistan to become highly sceptical about the ongoing military operations. For instance, the results of a survey of rural and urban Pakistanis conducted in October 2008 indicated that a majority of Pakistanis did not support the Pakistani Army fighting in FATA and NWFP (Fair & Jones, 2009, p. 179). The need for this change was realised at the highest level, that is, by General Kiyani. Although he took a long time to formally kickstart the process of change in 2011, this time delay can be attributed to the overstretched commitment of the Pakistani Army, being a coalition partner in GWOT, fighting against the Taliban in FATA. Kiyani acted as a "military meteorologist" for the Pakistani Army by appreciating the need for change and systematically analysing various learning cycles to aptly forecast the changes/improvements required in the military culture of fighting insurgencies to best prepare for the future challenges. He gathered a set of officers around him who could think critically. The net result of this effort was the SCW doctrine in 2013, which brought greater clarity, flexibility and operational improvement in the SCO of the Pakistani Army (Adnan 2019, Interview, 14 January). As Mumford & Reis maintain:

Any military organisation wishing to achieve greater flexibility, improvement in tactical and operational effectiveness, or strategic clarity, therefore must seek individual leaders with the capacity to think critically about the nature of their experiences and extract pertinent lessons for the purpose of theorization and doctrine.

(2014, pp. 7–8)

The SCW doctrine (2013) is the synthesis of all the learning cycles of the Pakistani Army as it addresses the performance gaps identified in all these cycles since 1947. The doctrine establishes the nature of the counterinsurgency problem in its broader context followed by addressing the causes of the insurgency whilst conspicuously highlighting the conflict environment in Balochistan. The SCW doctrine (2013) provides elaborated guidance in broadly three dimensions, namely security and military operations, political and legal aspects, and finally the CIMIC of the sub-conventional warfighting. These three strands correspond to the three pillars of Kilcullen's Counterinsurgency Model. Generally, the officers of the Pakistani Army maintain that the performance gaps of the Pakistani Army identified over the long period since partition are mostly addressed in the SCW doctrine (2013) owing to the lessons learned the hard way in various counterinsurgency campaigns (Qadeer 2022, Interview, 08 December; Ahsan 2019, Interview, 16 January; Kazmi 2018, Interview, 19 December). The institutionalisation of the learning took place through various training regimes introduced into the army, which are still in vogue.

The process of learning, the institutionalisation of the learning and the themes enshrined in the SCW doctrine, as elaborated in the previous and current chapter, contribute to the counterinsurgency approach of the modern-day Pakistani Army. That is how the two Chapters (3 and 4), deal with the sub research question of the study: *What is the Doctrinal Approach of the Pakistani Army towards Counterinsurgency?*

As Catignani maintains:

New doctrine may be written (or rediscovered), but may continue to be ignored by the organization's personnel. Lessons learned processes and structures may be created or refined, but can still remain predicated on improving the organization's short-term operational performance .... rather than on reconceptualising new ways in which personnel can understand, prepare for and, consequently, conduct COIN warfare on the ground.

(2012, p. 536)

These two chapters set a firm foundation on which the later empirical chapters of the book, after assessing the threats to CPEC in real terms (Chapter 5), will be based to evaluate two aspects. Firstly, the extent to which the Pakistani Army adheres to the doctrinal approach, as enshrined in the SCW document, by incorporating it into the COIN strategy in Balochistan. Secondly, the efficacy of the COIN strategy of the Pakistani Army in fighting insurgency in Balochistan after 2013 concerning the safety of CPEC. This will address Catignani's observation.

Notes

1 During the entire tenure of General Raheel Sharif (immediate successor of Kiyani) as COAS (2013–16), the Pakistani Army was in a process of transformation from a conventional force to a more COIN oriented force in the light of the SCW doctrine (2013). This transformation took a long time. This was obviously due to the enormous size of the Pakistani Army which is the 6th biggest standing army in the world with a total strength of 1.15 million troops (Mizokami, 2019) including the reserves and the paramilitary forces which operate under the operational command of the army (Jaffer 2019, Interview, 17 January; Muzammil, 2018, Interview, 27 December). Although General Raheel Sharif took several measures like formalised the COIN training by including it in the curricula of the Pakistani Army for the first time in its history, introduced PIT, laid emphasis on the Information Operations (IOs) as per the SCW doctrine (2013) but still the significant change was not visible in the COIN approach during the operations in FATA and Balochistan during his tenure. The practical manifestation of the SCW doctrine (2013) took considerable time (Nadeem 2018, Interview, 13 December; Adnan 2019, Interview, 14 January; Fahad 2022, Interview, 29 January).

2 It is beyond the scope of the book and space limitation of this section to cover all the schools of thought/ scholarly literature on military innovation which is massive. It is not even possible to cover the nuanced discussion of the main works on military innovation. Only the literature which is relevant and helps to understand the innovation concerning the counterinsurgency operational praxis in the Pakistan Army is discussed.

3 "By 'mavericks' Posen means officers with unconventional ideas who are willing to cooperate with civilians to reshape the military" (Grissom, 2006, p. 909, foot note 20).

4 Defence budget was increased to $6.41bn (£4.08bn) from $5.14bn (£3.27bn) in the 2010–11 budget approved by parliament (BBC News, 2010).

5 See Theo G. Farrell and Terry Terriff, "The Sources of Military Change", in idem, Sources of Military Change, 8–10. See also specifically Theo G. Farrell, "World Culture and the Irish Army, 1922–1942", in Farrell and Terriff, Sources of Military Change, 69–90.

6 See for example Catignani (2014, p. 32) "Coping with Knowledge: Organizational Learning in the British Army". This has been the case with the British Army where a large number of army personnel (interviewees) confessed to the author that they had to "relearn lesson learned" and "reinvent the wheel" during their deployment at Helmand Province of Afghanistan.

References

Abrahamsson, B. (1972). *Military Professionalization and Political Power*. London: Sage.

Ahmed Ashfaq. (2016) 'Pakistan: Two years on, army says fight not yet over', *Gulf News,* 26 June. Available at: https://gulfnews.com/world/asia/pakistan/pakistan-two-years-on-army-says-fight-not-yet-over-1.1852707 (Accessed: 19 March 2022).

Army Council (1937). *Duties in Aid of the Civil Power (WO 279/469)*. London: War Office Publication.

Avant, D. (1993). 'The Institutional Sources of Military Doctrine – Hegemons in Peripheral Wars', *International Studies Quarterly,* 37 (4) (Dec. 1993), pp. 409–430.

Avant, D. (1994). *Political Institutions and Military Change: Lessons from Peripheral Wars*. Ithaca, NY: Cornell UP.

Banerjee, D. (2009). 'The Indian Army's Counterinsurgency Doctrine', in Sumit, G. & David, P. F. (eds.) *India and Counterinsurgency: Lessons learned*. Abingdon, Oxon: Routledge, pp. 189–206.

Barnett, C. (1963). *The Sword bearers: Studies in Supreme Command in the First World War*. London: Eyre & Spottiswoode.

BBC News (2010) *Pakistan increases its defence budget*. Available at: https://www.bbc.co.uk/news/business-11391644 (Accessed: 02 December 2022).

Catignani, S. (2012). 'Getting COIN' at the Tactical Level in Afghanistan: Reassessing Counter-Insurgency Adaptation in the British Army', *Journal of Strategic Studies,* 35 (4), pp. 513–539. doi: 10.1080/01402390.2012.660625.

Catignani, S. (2014). 'Coping with Knowledge: Organizational Learning in the British Army?, *Journal of Strategic Studies,* 37 (1), pp. 30–64. doi: 10.1080/01402390.2013.776958.

Cohen, E. A. (2004). 'Change and Transformation in Military Affairs', *Journal of Strategic Studies,* 27 (3), (September), pp. 395–407.

Dawn (2017b) *Pakistan Army launches 'Operation Radd-ul-Fasaad' across the country*. Available at: https://www.dawn.com/news/1316332 (Accessed: 30 December 2022).

Dixit, K. C. (2010). 'Sub-Conventional Warfare Requirements, Impact and Way Ahead'. *Journal of Defence Studies,* Institute for Defence Studies and Analyses, New Delhi, 4 (1) (January), pp.120–134. Available at: https://idsa.in/jds/4_1_2010_Sub-ConventionalWarfareRequirementsImpactandWayAhead_kcdixit (Accessed: 02 April 2022).

Fair, C. C. (2014). *Fighting to the End: The Pakistan Army's Way of War*. New York, USA: Oxford University Press.

Fair, C. C. & Jones, S. G. (2009). 'Pakistan's War Within', *Survival,* 51 (6), pp. 161–188. doi: 10.1080/00396330903465204.

Fair, C. C. & Jones, S. G. (2010). *Counterinsurgency in Pakistan*. Santa Monica, CA: RAND Corporation.

Farrell, T. (1996). 'Figuring Out Fighting Organizations: The New Organizational Analysis in Strategic Studies', *Journal of Strategic Studies,* 19 (1), (Spring 1996), pp. 122–135.

Farrell, T. (1998). 'Culture and Military Power', *Review of International Studies,* 24 (3) (Fall 1998), pp. 407–416.

Farrell, T. (2002). 'Constructivist Security Studies: Portrait of a Research Program', *International Studies Review,* 4 (1) (spring), pp. 49–72.

Farrell, T. (2010). 'Improving in War: Military Adaptation and the British in Helmand Province, Afghanistan, 2006–2009', *Journal of Strategic Studies,* 33 (4), pp. 567–594. doi: 10.1080/01402390.2010.489712.

Farrell, T. (2020) 'Military Adaptation and Organisational Convergence in War: Insurgents and International Forces in Afghanistan', *Journal of Strategic Studies,* 45 (5), pp.1–25. doi: 10.1080/01402390.2020.1768371.

Farrell, T. & Terriff, T. (2002). *The Sources of Military Change: Culture, Politics, Technology.* Boulder, CO: Lynne Rienner.

Finer, S. (2002). *The Man on Horseback: The Role of the Military in Politics.* Piscataway, NJ: Transaction Publishers.

Grissom, A. (2006). 'The Future of Military Innovation Studies', *Journal of Strategic Studies,* 29 (5), pp. 905–934. doi: 10.1080/01402390600901067.

Hashim, A. (2013) 'Pakistan: A political timeline', *AlJazeera News,* 30 April. Available at: https://www.aljazeera.com/indepth/interactive/2012/01/20121181235768904.html (Accessed: 25 October 2022).

Huntington, S. P. (1957). *The Soldier and the State: The Theory and Politics of Civil-Military Relations.* Cambridge, MA: Belknap/Harvard.

ISPR. (2017). *Press Release No PR-87/2017.* 22 February. Available at: https://www.ispr. gov.pk/press-release-detail.php?id=3775 (Accessed: 19 June 2022).

Khan Iftikhar, A. (2018) 'New internal security policy aims at reconciliation with radicals willing to shun violence', *Dawn,* 31 May. Available at: https://www.dawn.com/news/1411174 (Accessed: 3 December 2022).

Kier, E. (1995). 'Culture and Military Doctrine – France Between the Wars', *International Security,* 19 (4), (Spring), pp.65–93.

Kier, E. (1997). *Imagining War: French and British Military Doctrine Between the Wars.* Princeton, NJ: Princeton UP.

Kilcullen, D. (2006) '*Three Pillars of Counterinsurgency*'. Available at: http://www.au.af. mil/au/awc/awcgate/uscoin/3pillars_of_counterinsurgency.pdf; https://tamilnation.org/armed_conflict/060928kilcullen.htm (Accessed: 22 January 2022).

Kitson, F. (1977). *Bunch of five.* London: Faber and Faber.

Lieven, A. (2011). *Pakistan: A Hard Country.* London: Allen Lane Publishers.

MacDonald, M. (2010) 'Pakistan army chief gets 3-year extension in office', Reuters, 22 July. Available at: https://www.reuters.com/article/pakistan-army-appointment-idUSSGE66L0RA20100722 (Accessed: 18 January 2022).

Mizokami, K. (2019) 'India vs. Pakistan: Who Wins in a War (And How Many Millions Could Die)?', *The National Interest,* 15 February. Available at: https://nationalinterest.org/blog/buzz/india-vs-pakistan-who-wins-war-and-how-many-millions-could-die-44827 (Accessed: 18 July 2022).

Mumford, A. & Reis, B. C. (2014). 'Constructing and Deconstructing Warrior-Scholars', in Mumford, A. & Reis, B. C. (eds.) *The Theory and Practice of Irregular warfare.* Oxon: Routledge, pp. 4–17.

Nagl, J. (2005). *Learning to Eat Soup with Knife.* Chicago, IL: University of Chicago Press.

Nawaz, S. (2011). *Learning by Doing – The Pakistan Army's Experience with Counterinsurgency.* Washington, DC: The Atlantic Council.

Nolan, V. (2012). *Military Leadership and Counterinsurgency: The British Army and Small War Strategy Since World War II.* London: I.B.Tauris.

Pakistani Army (1972). *Post-War Committee's Report on 1971 War in East Pakistan.* Rawalpindi: Chief of General Staff (CGS) Secretariat, GHQ.

Pakistani Army (1979). *Post-Conflict Evaluation Committee's Report on Balochistan-1979.* Rawalpindi: Chief of General Staff (CGS) Secretariat, GHQ.

Pakistani Army (2011). *Pakistan Army Doctrine (PAD) (Publication No. AP 1001E).* Rawalpindi: Doctrine & Evaluation Directorate, GHQ.

Pakistani Army. (2013). *Sub Conventional Warfare (SCW) Doctrine (Publication No. AP 2601E).* Rawalpindi: Doctrine & Evaluation Directorate, GHQ. Available at: https://www.scribd.com/document/474069737/Sub-Conventional-Warfare-Doctrine-pdf (10 August 2022)

Posen, B. (1984). *The Sources of Military Doctrine: France, Britain, and Germany Between the World Wars*. Ithaca, NY: Cornell UP.

Puri, S. (2012). *Pakistan's War on Terrorism*. Abingdon, Oxon: Routledge.

Rashid, A. (2008) 'Pakistan's Worrisome Pullback', *The Washington Post*, 06 June, p. A19.

Rosen, S. P. (1991). *Winning the Next War: Innovation and the Modern Military*. Ithaca, NY: Cornell UP.

Shaikh, A. (2014) 'A leaf from history: Reclaiming Balochistan, peacefully', *Dawn,* 05 October. Available at: https://www.dawn.com/news/1135570 (Accessed: 20 February 2022)

Siddiqui, N. & Akbar, A. (2017) 'Major among 4 Pakistan Army personnel martyred during operation in Upper Dir', *Dawn,* 09 August. Available at: https://www.dawn.com/news/1350539 (Accessed: 30 December 2022).

The Express Tribune (2013) *New doctrine: Army identifies 'homegrown militancy' as biggest threat*. Available at https://tribune.com.pk/story/488362/new-doctrine-army-identifies-homegrown-militancy-as-biggest-threat/ (Accessed: 25 March 2022).

Thompson, R. (1966). *Defeating Communist Insurgency*. Reprint, St. Petersburg, FL: Hailer Publishing, 2005.

Ucko, D. (2009). *The New Counterinsurgency Era: Transforming the U.S. Military for Modern Wars*. Washington, D.C.: Georgetown University Press.

Ullah, K. A. (2017) *Gains of Radd-ul-Fassad*. Available at: http://issi.org.pk/wp-content/uploads/2017/08/Final_IB_Asad_dated_18-8-2017.pdf (Accessed: 18 March 2022).

Zisk, K. (1993). *Engaging the Enemy: Organization Theory and Soviet Military Innovation 1955–1991*. Princeton, NJ: Princeton UP.

5   Security Threats to China Pakistan Economic Corridor (CPEC) in Balochistan

"Anywhere [in Balochistan] the Chinese are working will be perceived as a CPEC project and could hence be subject to attack".

Kaiser Bengali[1]
(2017, cited in International Crisis Group's Report, 2018, p. 21)

Introduction

Kilcullen's Three Pillars of Counterinsurgency Model (3PCM), as described in Chapter 2, has two parts. The first part pertains to the elaboration of the "conflict ecosystem" (Figure 2.1, Chapter 2) that forms the environment for twenty-first-century counterinsurgency operations. In the case of the Balochistan Conflict Ecosystem, many independent but interlinked actors exist. These include insurgent/terrorist organisations such as the Baloch Liberation Army (BLA), Baloch Republican Army (BRA), and the Balochistan Liberation Front (BLF). This also includes the sectarian militant organisations like Lashkar-e-Jhangvi Al-Alami (LeJ), Jamaat-ul-Ahrar (JuA), and Sipah-e-Sahaba (SSP). The new state of the environment (on-going fifth round of insurgency) has drawn new actors like the Islamic State of Khorasan (ISKP) as shown in Figure 2.2 (Chapter 2). These insurgents, sectarian militants, and global jihadists are foreign-funded, and they get their supply of weapons, equipment, ammunition, and hard-core guerrilla training from outside because of the porous border of Balochistan with Afghanistan and Iran (see Appendix 1 for the international border alignment)[2]. All these actors in the prevailing Balochistan "conflict ecosystem" pose a potent threat to the CPEC in the province. Therefore, it is imperative to first assess carefully the diversity and quantum of the threat to the CPEC in Balochistan before evaluating the counterinsurgency (COIN) strategy (2013–2022) of the Pakistani Military using the 3PCM.

The China Pakistan Economic Corridor (CPEC) has been subject to grave security threats since its inception especially in the restive southwestern province of Balochistan, where major elements of strategic importance to CPEC like the 1.6 billion USD Gwadar deep water port are located. Security issues are one of the most significant challenges to CPEC (Chaziza, 2016, p. 151; Ibrar et al., 2016; Kulkarni, 2022) and a matter of principal concern both for Beijing and Islamabad (Basit, 2019, p. 705; Kulkarni, 2022). Small (2015, p. 174) argues that one of the

DOI: 10.4324/9781003413905-5

primary concerns for Beijing is the increasing threat of terrorism within Pakistan. Beijing's Security concerns over the CPEC were clearly expressed to Prime Minister (PM) Shehbaz Sharif on the latter's official visit to China in November 2022. A Chinese Foreign Ministry statement said, "President Xi expressed his great concern about the safety of Chinese nationals in Pakistan and conveyed his hope that Pakistan will provide a reliable and safe environment for Chinese institutions and personnel working on cooperation projects [CPEC] there" (Gul, 2022).

In Balochistan, the fifth round of secessionist insurgency started in 2006, and a number of violent militant sectarian and secessionist organisations also exist in the province. The autonomy-seeking Baloch have historically viewed the Federal Government as a usurper of their rights. In more recent years they have accused the Federal Government of colonising the province with the assistance and collaboration of the Chinese (Brahamdagh Bugti, 2015, cited by Deutsche Welle (DW), 2015). Since the formal commencement of CPEC, a large number of attacks have been carried out by these secessionist insurgent organisations, as well as by the Islamic State and sectarian militant organisations. These attacks have been against CPEC infrastructure, Chinese nationals working on CPEC projects along with the local workers, law enforcement agencies (LEAs), and ethnic/sectarian minorities in the province and a couple of attacks on Chinese nationals in Karachi as well. For instance, a day after the suicide bombing attack on Karachi University's Confucius Institute in 2022, a video circulated on social media. In this video message, a masked BLA commander (speaking in English) warned the Chinese leadership to quit Balochistan or "witness retaliation from Baloch sons and daughters" and said:

> China, you came here without our consent, supported our enemies, [and] helped Pakistani military in wiping [out] our villages, but now it's our turn. The Baloch Liberation Army (BLA) guarantees you that China Pakistan Economic Corridor will fail miserably on Baloch land.
>
> (Jamal, 2022)

Moreover, in recent years, the emergence of the ISKP[3] in the province and its collaboration/coordination with the existing insurgent groups and the militant sectarian organisations has added yet another dimension to the existing complex "conflict ecosystem" of Balochistan. This has contributed further to the worsening of the security milieu in Balochistan (Ahsan 2019, Interview, 16 January; Durrani 2022, Interview, 11 August). The operational activities of ISKP against the CPEC have given the regional insurgency in Balochistan an international dimension.

The security threats to CPEC have been felt at the governmental, diplomatic, and military levels on both sides (Gul, 2022). Statements have been issued from the Chinese side in the aftermath of militant attacks urging Pakistan to enhance security. The impatience and sensitivity of the Chinese side to the security of CPEC is evident from its rush to include the Pakistani Military as a stakeholder in the overall security design of CPEC (Arduino, 2017). These Chinese statements, from various quarters, have mixed tones and messages. Sometimes it is a harsh reprimand made publicly, and on other occasions, the statement may praise Pakistani efforts to curb militancy and stress the need for greater cooperation for providing fool proof security to CPEC and Chinese nationals. At other times they may just be a warning against future terrorist attacks. For example, in 2015 General Fan Changlong, the Vice Chairman of the Central Military Commission of China, during a visit to the Pakistani Army's General Headquarters argued: "We [the Chinese] look forward to close cooperation with Pakistan to ensure proper management and security of the CPEC" (Syed Baqir, 2015). Furthermore, the Vice Chairman maintained that his government would meet the Pakistani Army's needs for CPEC protection (Syed Baqir, 2015). Sun Weidong, the Chinese Ambassador to Pakistan, in 2016 stated that the "construction and safety of CPEC would have to go side by side". He further observed, "We look forward to creating a safe and sound environment for CPEC along with the Pakistani side" (Syed Baqir, 2016). Another press release by the Chinese embassy in Islamabad, issued on December 8, 2017, reads: ".... We appreciate Pakistan has attached much importance to the security of the Chinese institutions and personnel" (Pakistan Today, 2017). After the militant attack (claimed by the BLA) on the Chinese consulate in Karachi on 23rd November 2018, the Chinese Embassy in Islamabad issued a statement, "We believe that the Pakistani side is able to ensure the safety of Chinese institutions and personnel in Pakistan ...... Any attempt to undermine China-Pakistan relationship is doomed to fail" (Chandran, 2018).

More recently, in April 2022, a female BLA militant carried out a suicide bombing near Karachi University's Confucius Institute, killing three Chinese nationals (language teachers), and their Pakistani van driver (Saeed, 2022). Following the attack, the Chinese foreign ministry issued a strong statement of condemnation. "China expresses its strong condemnation and great indignation at this major terrorist attack", tweeted Zhao Lijian, the deputy director of the Chinese foreign ministry. He urged Pakistani authorities to "deal with the aftermath" and "resolutely fight against terrorist organisations involved in the case" (BBC, 2022).

Correspondingly the statements by the Pakistani side in the aftermath of any insurgent incident aimed at sabotaging the CPEC, are always made unapologetically together with a firm reiteration to provide security to the CPEC/Chinese nationals and workers. General Qamar Bajwa the former Chief of the Army Staff (COAS) of the Pakistani Army and his predecessor General Raheel Sharif avowed through press statements in 2018 and 2016 respectively that the Pakistani Army would ensure security of CPEC at all costs, and that it was aware of the hostilities against the project (Dawn, 2018; Haider, 2016). On the political side, both the former prime ministers of Pakistan, Imran Khan and his predecessor Nawaz Sharif, issued statements about providing security to the CPEC. Lately, in the wake of the suicide attack near Karachi University's Confucius Institute, senior Pakistani officials "scrambled to appease China" (BBC, 2022). PM Shehbaz Sharif visited the Chinese embassy in Islamabad and presented officials with a handwritten letter reaffirming Pakistan's commitment to eliminating all militants and terrorists from its soil. "We won't rest until the culprits are hunted down and given exemplary punishment", he wrote (BBC, 2022). Presenting the handwritten letter by the PM to condemn the incident and show resolve to end insurgents' attacks on the Chinese was an unprecedented move which clearly showed prevailing anxiety amongst the top officials about the security situation concerning CPEC and the Chinese nationals working on various projects in the country. Pakistan's foreign office spokesperson, Asim Iftikhar, described the incident as "reprehensible" and a "direct attack" on Pakistan-China relations. Former Prime Minister Imran Khan also commented on the attack, stating it was "yet another attack with a specific agenda of trying to undermine Pakistan-China strategic relationship" (BBC, 2022).

These statements by both the Pakistan and Chinese political and military leadership establish a few facts: a) security-wise CPEC is a high-risk project in Balochistan; b) both sides are conscious of the security issues pertaining to CPEC and display high resolve in public to secure the project and; c) the success of the project depends on the stabilisation of the security situation in the country in general and Balochistan in particular as a high-security threat to the project lies in the province.

This chapter aims to unfold the various dimensions of the existing security threat spectrum to CPEC in Balochistan in terms of highlighting the insurgent organisations, militant sectarian organisations, and ISKP's agendas, ideologies, and motives. It also details the links between these groups. The chapter sets out how these groups' aims, objectives, and tactics pose threats to the CPEC. This chapter will also set out how these groups challenge the counterinsurgency doctrine that the Pakistani Army put in place in Balochistan against the backdrop of providing safety to the CPEC. The efficacy of the above-mentioned doctrine and the overall counterinsurgency policy of the Pakistani Army in Balochistan will be assessed in the next chapter.

Threat Spectrum for CPEC in Balochistan

The threat spectrum for CPEC in Balochistan can broadly be divided between internal and external threats. The chapter will focus more on the internal threats. The internal threats are very potent and have very complex dynamics (Nadeem 2018, Interview, 13 December; Adnan 2019, Interview, 14 January; Durrani 2022, Interview, 11 August). As Basit aptly puts it, "the militant groups that have targeted Chinese interests in [Balochistan] Pakistan have had a religious or separatist agenda" (2019, pp. 704–705). The internal threat consists of the secessionist insurgent organisations, militant sectarian organisations, many of whom have been declared internationally as terrorist organisations, and the emerging threat of the ISKP. ISKP in Balochistan is a comparatively new phenomenon. The external threat originates from the porous borders of Balochistan with Afghanistan and Iran which is used by the militants and insurgents in fleeing from the operations of the LEAs as well as for logistics and arms/ammunition supplies, training camps, bases, and a launch pad for carrying out terrorist attacks in the province. Besides this, Islamabad has blamed India for igniting the insurgent activities in Balochistan for fulfilling its ulterior goal of sabotaging the CPEC (Iqbal, 2018, p. 207; Zeb, 2019, p. 169). For instance,

Pakistan's Ministry of Foreign Affairs (MOFA) press release (no. 73/22) issued on 10 February 2022 states,

> We also strongly reject India's persistent propaganda against China-Pakistan Economic Corridor (CPEC). Pakistan has shared irrefutable evidence of India's sinister campaign to sabotage CPEC through its dossiers released in 2020 and 2021. There is strong evidence of Indian involvement in recent sinister attempts to stir up unrest in Balochistan by supporting anti-state elements.
>
> (MOFA, 2022)

In addition, the funding from Saudi Arabia and Qatar to the religious seminaries (madrassahs) in Balochistan (Siddiqa, 2018) and the financing from UAE and Oman to Baloch insurgents are other external threats to CPEC (Dorsey, 2019, pp. 148–151). All these threats are discussed, in detail and thematically, below.

Internal Threats

Baloch Secessionist Insurgency/Organisations

One of the lethal attacks by the Baloch secessionist organisation against the Pakistani Military was conducted on 18th April 2019 when around 15 to 20 assailants disguised in Frontier Corps (FC) uniform stopped a bus on the Makran coastal highway which connects Gwadar to Karachi. The militants offloaded 16 passengers after checking their identity cards. Fourteen passengers were shot dead by the militants while two managed to escape safely (Khaleej Times, 2019). The slain passengers included ten personnel of the Pakistani Navy, three of the Pakistani Air Force, and one from the Pakistani Coast Guards (Associated Press of Pakistan (APP), 2019). The foreign minister of Pakistan confirmed that the responsibility of the attack was claimed by Baloch Raaji-Aajohi e-Sangar (BRAS), an alliance of three insurgent Baloch organisations: BLF, BLA, and the Baloch Republican Guard (BRG) (APP, 2019; The Balochistan Post, 2018b). It is essential to highlight that the Makran coastal highway is a well-developed and fully functional road infrastructure, with over half a dozen check posts of various LEAs, on the central and eastern route alignment of the CPEC. This attack served as a clear message of the threat to CPEC and the Chinese interests in Balochistan.

Terrorist incidents targeting CPEC have occurred at regular intervals over the past few years in Balochistan. Sometimes the victims are LEAs personnel and sometimes Chinese nationals or locals working on the CPEC linked projects, at other times the CPEC infrastructure is targeted. For example, in August 2018, the Baloch insurgents BLA carried out their first-ever suicide attack against Chinese nationals in Balochistan by targeting a bus carrying Chinese engineers under the security of the Pakistani Military (Wu, 2018). Likewise, in November 2018, the insurgents of BLA targeted the Chinese consulate in Karachi in Sindh province (BBC News, 2018b). This was the first-ever terrorist attack of its type conducted by any Baloch secessionist organisation outside Balochistan. More recently, in April 2022, a suicide attack by BLA near Karachi University's Confucius Institute killed three Chinese nationals (BBC, 2022). This attack, like many others targeting Chinese nationals and interests, was an eye-opener because it raised the question as to why Baloch secessionist or-ganisations would target Chinese personnel and interests in Pakistan. The answer to this question lies in the Chinese initiative and investment of CPEC.

The Baloch separatist movement has been going on for a long time and the in-surgency is in its fifth round. The secessionist agenda of the Baloch people is very much the result of, as aptly observed by Harrison, "the conviction that Baluchistan contains vast, untapped natural wealth is central to the separatist creed .... " (1978, p. 144). This observation is pertinent in today's Balochistan as well. The separatist agenda is not only because of the existence of large amounts of natural resources in the province but is equally a result of the sense of deprivation, usurpation, and exploitation of the natural resources of Balochistan by the federation. Both the Baloch insurgents and nationalists believe that China is a "crime partner" with the Government of Pakistan in exploiting and plundering the natural resources of Balo-chistan (The Balochistan Post, 2018a). Shafqat, a Baloch academic at Balochistan University, Quetta, argues that China was even involved in exploiting the natural resources of Balochistan far earlier than CPEC (2018, Interview, 24 December). For instance, the China Metallurgical Group Cooperation won the tender and ex-tracted gold and copper from the Saindak mines in Balochistan during the 1990s. The Baloch nationalists view such projects as the exploitation of natural resources. As Adeney (2012, p. 544) has discussed, the central government has historically benefited from natural resource extraction. Ahsan (2019, Interview, 16 January) argues that Baloch nationalists staunchly believe that CPEC has just formalised the joint "plundering" of the natural resources of Balochistan and earning revenue from various projects by the Chinese and the Pakistani Governments. This percep-tion persists despite changes made under the 18th Amendment in the Constitution of Pakistan. The reason is that the 18th Amendment changed the formula, but this has failed to assuage Baloch separatists because the measures relating to revenue sharing were not implemented (Adeney, 2012, p. 552).

This anti-Chinese sentiment of the Baloch insurgents is substantiated by a video message, widely circulated on social media, of Raziq Baloch and Raees Baloch the BLA Fedayeen (guerrillas) who attacked the Chinese consulate in Karachi. They attacked China and Pakistan by stating:

> China along with "occupying" Pakistani [military] in Balochistan not only plundering Baloch resources but also using Baloch land, particularly Gwadar and other coastal areas to further their evil military designs .... we want to reit-erate to China that they cease their exploitative and evil plans and projects [in Balochistan] on an immediate basis, or they will continue to face the Baloch youth. Our attacks [against Chinese] will be further intensified and frequent.
>
> (Khan Wajahat, 2018)

A similar video message threatening the Chinese and Pakistani governments to abandon CPEC in Balochistan was once again circulated by the BLA in April 2022, following the suicide attack on Karachi University's Confucius Institute, as mentioned earlier in the chapter.

It is essential to take an in-depth look at each of the main insurgent groups in Balochistan, in turn, to fully comprehend the threat spectrum for the CPEC in Balochistan posed by these secessionist organisations.

Balochistan Liberation Army (BLA)

The Balochistan Liberation Army (BLA), also known as the Baloch Liberation Army, is an ethno-secessionist insurgent organisation active in Balochistan since it was established in 2000. The organisation has its bases in the border regions of Afghanistan and Iran. The BLA has been declared as a proscribed terrorist organisation both in the UK (Home Office: Proscribed Terrorist Organisations, 2021) and Pakistan (NACTA, 2019). In July 2019, the BLA was banned by the United States and branded as a global terrorist group just ahead of the former Prime Minister Imran Khan's first state visit to Washington DC (Farmer, 2019b). As a result of this decision, the United States authorities were able to freeze financial assets linked with the BLA in the United States.

The leadership of the organisation views CPEC as an example of the economic colonisation of the Baloch people and the plundering of the natural resources of the province. The organisation is the largest and most active in the province, and it is estimated to have the manpower of 1000–2000 trained insurgents (Ahsan 2019, Interview, 16 January). The BLA procure their weapons from Afghanistan and Iran. Moreover, the financial donations from the Baloch diaspora, especially in the Middle East and the Persian Gulf countries, helps the organisation to buy ammunition and weapons from the black market (Khan, 2009). The organisation claims that it receives handsome donations from the Baloch people as the cause of an independent Balochistan is very popular amongst the Baloch nation (BLA Media, cited in Stanford CISAC, 2019). Apart from these sources of finance, some experts believe that the smuggling of drugs [opium/heroin] through Balochistan also generates funds for the BLA and other Baloch armed groups (Grare, 2006, p. 10; Lieven, 2017, p. 179). The BLA works in collaboration with the other Baloch secessionist organisations as well as the ISKP. The BLA dominates the Zhob, Sibbi, Musa Khel, Nokundi, Dalbandin, and Panjgoor areas of Balochistan. Moreover, the organisation is very active in Makran division. It is the most active, organised and established secessionist organisation in Pakistan with its sleeper cells in almost all major cities of Balochistan (Durrani 2022, Interview, 11 August). It is supported by sympathy for the free Balochistan cause from a large number of Baloch people in the country and abroad including Iran and Afghanistan (Zaka 2019, Interview, 03 January).

In recent years, the BLA has increased its sphere of insurgent activities to target Chinese nationals and investments outside Balochistan as well, particularly in the port city of Karachi, which shares the land route with Balochistan's Hub district (Saeed, 2022). Notable incidents include a) a November 23, 2018 attack on the Chinese consulate in Karachi by three BLA-affiliated suicide bombers (Janjua & Shams, 2018) and b) a more recent suicide bombing carried out by a female BLA militant named Shari Baloch in April 2022 near Karachi University's Confucius Institute. This attack resulted in the killing of three Chinese nationals and their Pakistani van driver (BBC, 2022).

Ideology and Goals

The BLA appeared as an insurgent organisation against the Pakistani state for protecting the rights of the Baloch nation by resorting to the armed struggle for independent Balochistan. It believes that Balochistan has been subjected to discriminatory and violent policies of successive federal governments for decades (Iaccino, 2016). The group seeks to rid Balochistan of foreign influence, specifically from the Chinese and Pakistani Government (Stanford CISAC, 2019)

Leadership

The BLA does not divulge the name of its leaders, but it is believed that the leadership cadre of the organisation consists of Bugti and Marri tribe members (Stanford CISAC, 2019). The government of Pakistan alleges that Hyrbyair Marri (the son of late Baloch nationalist leader Khair Bakhsh Marri) is the leader of the BLA. His second-in-command was Aslam Baloch Achu (the mastermind and planner of the attack on the Chinese Consulate in Karachi) who took over the leadership of the organisation in Pakistan and Afghanistan, but he was killed in a suicide attack in Afghanistan (Tolo News, 2018). Hyrbyair went into exile in 2000 once the Government of Pakistan started operations against the Marri tribe. He was granted asylum by the British Government in 2011 (Shah Murtaza, 2011).

Likely Targets of BLA

At present the organisation is involved in targeting the LEAs, attacking Chinese nationals, and local civilians working on the CPEC, sabotaging CPEC infrastructure and suicide bombings in Balochistan.

Balochistan Republican Army (BRA)

The BRA is an ethno-secessionist insurgent organisation that is alleged to be the militant wing of the Baloch Republican Party (Samad, Singh, & Samad, 2015, p. 124). It was founded in 2006 as a consequence of increasing resentment in Balochistan against the Pakistani state's central control over the resources of Balochistan. Brahamdagh Bugti (grandson of the slain Akber Bugti who was killed by the Pakistani Army in 2006) is the leader of the BRA. Brahamdagh Bugti stayed four years in Kabul (2006–10) and managed the insurgent activities of the organisation (Samad, 2014, p. 300). He is accused of killing civilians, sabotaging infrastructure/government installations and attacking the LEAs in the province. In 2010, Brahamdagh Bugti left for Switzerland to seek asylum. This organisation is also busy fighting with the Pakistani state. The government of Pakistan declared BRA a terrorist organisation in 2010 (NACTA, 2019). The organisation operates in Quetta, Mastung, Kalat, Noshki, and Chagi districts of Balochistan (Abdul Basit, 2016, p. 234).

Ideology and Goals

The BRA is an ethno-nationalist separatist organisation working for the creation of Balochistan as an independent state (Stanford University, 2015a). The organisa-tion views any foreign investment, including the CPEC initiative, "as an attempt to plunder the rich natural resources belonging to the Baloch Nation" (Badini (a for-mer Baloch insurgent) 2019, Interview, 15 January). Therefore, the BRA is against any such investment in Balochistan (Stanford University, 2015a).

Likely Targets and Tactics of BRA

Like the BLA, the BRA also targets the LEAs, road/rail infrastructures, public transport, natural gas pipelines, the civilian population, foreigners (mainly Chinese) in the province. The insurgents use small arms such as mortars and landmines. They also resort to the planting of IEDs at public places and on the likely routes of the military convoys.

Balochistan Liberation Front (BLF)

The BLF is one of the oldest Baloch insurgent organisations, established in 1964 in Damascus by Juma Khan Marri (Stanford University, 2015b). In 1968, the organi-sation joined the Baloch insurgency in Iran against the Iranian government. During this revolt, the then Iraqi government provided full materialistic and operational support along with weapons and ammunition to BLF in an effort to destabilise Iran (Stanford University, 2015b). After the termination of the conflict in Iran, the BLF initiated an insurgency against Pakistan to get the independence of Balochistan on the lines of the recently liberated East Pakistan (Bangladesh). This fourth round of the Balochistan insurgency is also known as the Independent Movement of Balochistan of 1973–1978. During this insurgency, Iraq was supplying weapons and ammunition to the BLF and other Balochi militant groups which were seized by the Pakistani forces after raiding the Iraqi embassy in Islamabad (Mashkoor, 2017, p. 17). The fourth round of insurgency came to an end in 1978 after the killing a large number of BLF activists and flushing out the remaining ones to Afghanistan. The activities of the BLF from 1977–2004 are uncorroborated (Stanford Univer-sity, 2015b), but it resurfaced in 2004 when it was responsible for the killing of three Chinese workers in Balochistan. Right after the incident, Dr Allah Nazar, the head of the BLF, accepted responsibility for the killings by the organisation and made a public statement that he had taken over the command of the BLF in 2003. The BLF expanded its activities to Awaran, Makran, and later on, Quetta appeared to be its hub of activities (Abdul Basit, 2016, p. 235).

Likely Targets

Since 2012, the BLF continued its attacks on LEAs and military personnel, journalists, foreign workers, and local manpower working on CPEC linked projects. For example, in April 2015, the BLF killed 20 workers working on CPEC projects in Balochistan (BBC News, 2015). Shortly after the attack, Nazar claimed that the slain individuals were targeted for being members of the FWO (Zurutuza, 2015).

Ideology and Goals

The BLF is an ethno-nationalist separatist organisation working for the creation of Balochistan as an independent state. The organisation openly denounces any foreign investment from China in the province, and it suspects that China is assisting the Pakistani Military to suppress the insurgency in Balochistan. In 2016, Nazar warned of more attacks against the CPEC projects in Pakistan (Hashim, 2016). In November 2018 after the Chinese consulate attack, Nazar proclaimed that "the attack on Chinese consulate by BLA was a natural reaction of the Baloch nation against Pakistani aggression and Chinese imperialistic ambitions" (The Balochistan Post, 2018a). He further added, "China is a crime partner with Pakistan in Baloch genocide to fulfil its imperialist aspirations and to tilt the balance of power in its favour in this region while employing Balochistan's strategic position and exploiting its natural resources" (The Balochistan Post, 2018a). More recently, in February 2023, Dr. Allah Nazar Baloch severely criticised Pakistan's judiciary, parliament, and media along with the Pakistani Army in his series of Tweets and accused them of a conspiracy against the Baloch nation. He also tweeted that there is little hope of justice for the Baloch people (News Intervention Bureau, 2023). These statements reflect the anti-Chinese and anti-Pakistani ideology of BLF and its goal to destabilise CPEC in Balochistan.

The above mentioned three insurgent organisations are the most active ones in Balochistan, sharing similar ideologies and relatively common goals. Owing to these similarities, the three organisations have recently entered into an operational collaboration and formed the BRAS. Little is known about BRAS except that it claimed responsibility for killing 14 military personnel on the Makran Coastal Highway in April 2019. BLA, BRA, and BLF along with ISKP account for ninety per cent of the insurgent incidents which happen in Balochistan, a senior police officer in Quetta has claimed (Zaka 2019, Interview, 03 January; Durrani 2022, Interview, 11 August). There are not many details available about other insurgent organisations such as LeB, BRG and BLUF in the province. These organisations are not very active, and they carry out activities on their own, without any collaboration with other organisations. BLA, BRA, and BLF are also in collaboration with sectarian militant organisations operating in Balochistan to further their agenda of derailing CPEC by dismantling the writ of the Pakistani state in Balochistan which will ultimately pave the way for a free Balochistan in their opinion.

Sectarianism in Balochistan

The exponential growth of sectarianism in Balochistan can be attributed to the military regimes' Islamisation policies for breaking the ethnolinguistic identities to change provincial demographics (Zaka 2019, Interview, 03 January). General Zia was the pioneer of this Islamisation strategy with particular focus on Balochistan and Khyber-Pakhtunkhwa (KPK). He considered this Islamisation strategy as an antidote to Baloch nationalism which, as per his thoughts, had fanned the flames of insurgency in Balochistan. This strategy led to the establishment of religious seminaries (madrassas) in the province. This strategy failed in the Baloch dominated areas of the province. The Islamisation strategy remained an essential element of the Federal Government policy in Balochistan, even during Musharraf's regime despite his "enlightened moderation" rhetoric. During Musharraf's regime, the establishment of madrassas continued in Balochistan, through the Ministry of Religious Affairs, to penetrate deep into the Baloch ethnic areas (Grare, 2013, p. 11). The external funding from Saudi Arabia, Qatar, Oman, and UAE to the madrassas and some of the sectarian militant organisations in Balochistan have also played a significant role in fostering the sectarianism in the province. This aspect will be discussed later in the chapter under the heading of "External Threats to CPEC". From 2010 onwards, Islamic radicalisation gradually increased and hit its peak in Balochistan, led by Deobandi madrassa network. At this point, Islamisation in the province took a qualitative change. The province has become a hub of sectarian outfits such as Lashkar-e-Jhangvi Al-Alami (LeJ), Jamaat-ul-Ahrar (JuA), Sipah-e-Sahaba (SSP) now known as Ahle-Sunnat-wal Jamaat (ASWJ) after being officially banned in Pakistan (Jadoon, 2018, p. 43). The presence of these sectarian militant outfits in Balochistan is the result of the state policy of Islamisation pursued since the 1970s, an extensive network of Deobandi madrassas being funded by Saudi Arabia, Qatar and UAE, neighbouring Afghanistan and the Afghan refugee camps in Pakistan which had been the principal source of recruitment for the TTP and these organisations. Until 2010, the Baloch dominated areas, unlike the Pashtun dominated areas in the province, were resistant to radicalisation and mostly secular (Grare, 2013). After 2010, the situation changed. Balochistan has seen the highest number of enrolments in madrassas in Pakistan along with the rapid growth in the number of madrassas (Fair, 2014, p. 18). The sectarian organisations are also recruiting manpower from the Baloch dominated districts, mainly through madrassas, based on Deobandi ideology. At present, many top leaders of LeJ in Balochistan are ethnic Baloch (Grare, 2013; Adnan 2019, Interview, 14 January; Durrani 2022, Interview, 11 August).

Sectarian violence is rampant in Balochistan (Zeb, 2019, p. 146) which takes the form of targeting the Hazara community, a Persian speaking Shiite community based in Balochistan and northern Afghanistan. On the 12th April 2019, at least 20 people belonging to the Hazara community were killed in a blast in a vegetable market at the Hazarganji area of Quetta (Balochistan) (McKirdy & Saifi, 2019). LeJ claimed responsibility for the attack. "For years, the Hazaras have lived in a state of perpetual fear in Quetta, and promises made by successive governments have failed to ensure their safety" (Masood, 2019). Pakistan's National Commission for Human Rights (NCHR) reports that 509 members of the Hazara community were killed and 627 others have been injured since January 2012 (NCHR 2018, p. 5). Abdul Khaliq Hazara, chairman of the Hazara Democratic Party based in Quetta, puts the numbers even higher, claiming that more than 1500 Hazaras have been killed in various incidents of targeted killing in Balochistan in recent years (Radio Free Europe Radio Liberty, 2018). The killing of Shiites, mainly Hazaras, in Balochistan is not limited to the sectarian organisations in the province. "Over the past couple of years Islamic State militants have also targeted the Hazara community in Baluchistan [Balochistan]" (Yousafzai, 2018a). Other minorities, like Christians and Zikris have also been targeted (ACN, 2022). For example, on 2nd April 2018, four Christian were killed in Quetta by armed militants (Hashim, 2018). Earlier on 17th December 2017, eight Christians were killed, and dozens were wounded as a result of a suicide and gun attack on a church in Quetta during the Sunday service (Hashim, 2017).

Lashkar-e-Jhangvi Al-Alami (Army of Jhangvi – International)

LeJ, established in 1996, is one of the most lethal groups involved in the sectarian killing of Shiites in Pakistan (Jadoon, 2018, pp. 26–27). LeJ defected from Sipah-e-Sahaba (SSP), a Deobandi Sunni organisation, due to internal party rifts within the leadership and it aligned itself to the Taliban in Afghanistan. LeJ subscribes to the hardline Takfiri Deobandi school of Islam and considers Shiites apostates (Yousafzai, 2018a). The organisation has fuelled unprecedented sectarian violence in Pakistan, especially in Balochistan (Stanford CISAC, 2018a). Besides targeting and killing hundreds of Hazaras (Shiite community members) in Balochistan, LeJ also focuses on targeting other minorities like Christians as well as calling for the elimination of Western influence and interests in the region after 2001 (Jadoon, 2018, p. 40).

Amongst the Pakistani terrorist organisations, LeJ has been the most effective, active, and lethal pioneer of suicide attacks in Pakistan; kicking off this trend in 2003 by targeting a Shia religious procession. The organisation changed its name in 2009 from LeJ to LeJ Al-Almi. Adding the suffix Al-Almi (meaning International) depicts the change from a regional to a global orientation (intent to wage a broader Jihad beyond Pakistan) for the organisation. The organisation has been designated as a terrorist organisation by the United Nations (UN, 2015), Pakistan (NACTA, 2019), UK (Home Office, 2021), and USA (BBC News, 2003).

Ideology and Goals

LeJ is a staunch follower of the Wahhabi[4]-influenced version of the Deobandi movement (Stanford CISAC, 2018a). The organisation considers Shiites as infidels and is thus virulently anti-Shiite in its agenda where it seeks to establish a Sunni state in Pakistan (South Asia Terrorism Portal (SATP), 2017a).

Ahle-Sunnat-wal Jamaat (ASWJ) [People of the Tradition and the Congregation]

ASWJ, earlier known as Sipah-e-Sahaba, Pakistan (SSP), was established in 1985 as an anti-Shiite organisation. Since its establishment, SSP has been actively involved in the sectarian killing of the Shiite community across Pakistan. SSP, the pro-al-Qaeda group, was banned by the Government of Pakistan in 2002 for its links to the terrorist outfit/activities (Hasan, 2012). After the official ban by the Government of Pakistan, the organisation was renamed in 2003 as Millat-e-Islamia Pakistan and renamed again as ASWJ in June 2008 and re-established as a political party (Hasan, 2009; SATP, 2017b). ASWJ was again banned by the Government of Pakistan for its "concerns in terrorism" in February 2012 (Hasan, 2012). However, the organisation remained active in targeting Shiite and other minorities throughout Pakistan, particularly in Balochistan after the ban. In June 2018, in a surprising move, the Government of Pakistan lifted the ban on ASWJ and ordered the unfreezing of the assets of its chief (Chaudhry, 2018). ASWJ is very active in killing Hazaras in Quetta in collaboration with LeJ and other militant sectarian organisations in the province of Balochistan.

Islamic State of Khorasan (ISKP), the Emerging Security Threat to CPEC in Balochistan

Jihadist Insurgency and terrorism struck Pakistan in its worst form over the last two decades especially after the commencement of the Global War on Terror (GWOT). There are several reasons for it. Foremost is the geographical location of the country, in terms of sharing a long porous border with the ever-turbulent Afghanistan, and its role in fomenting the Afghan Jihad against the erstwhile Soviet Union at the behest of the CIA. After the fall of the Soviet Union and its withdrawal from Afghanistan, the latter became the breeding ground of terrorists from across the globe. Though the GWOT ended the Taliban regime in Afghanistan, other terrorist organisations like Al Qaeda and Tahreek-i-Taliban, Pakistan (TTP) surfaced in the Afghanistan-Pakistan (Af-Pak) border region. The Pakistani Military fought a long war against jihadist insurgents in the country. Although the war against terrorists in Pakistan gained considerable success, most of the time the porous border between Afghanistan and Pakistan enabled the leadership to flee to Afghanistan in the aftermath of operations (Adnan 2019, Interview, 14 January). The defection and changing alliances of various terrorist outfits gave rise to new terrorist brands in the region. For example, the defectionist elements of Al Qaeda and TTP joined the rank and file of ISIS and laid the foundation of ISKP in the region (Jadoon, 2018, p. 1). Establishing Wilayat Khorasan (Khorasan Province) is a crucial part of the Islamic State's "quixotic plan" of the global Islamic Caliphate (Stanford CISAC, 2018b). The Khorasan region spans the Af-Pak region and extends to parts of Central Asia, China, and Iran (Jadoon, 2018, p. 6). The Khorasan region in general and the Af-Pak region, in particular, is swamped with a variety of militant groups that have overlapping and competing ideologies and agendas. This characteristic of the region presents opportunities for the ISKP brand, as it creates the potential to generate alliances.

ISKP currently poses a serious threat to the region, especially Pakistan where it has established a strong foothold in the province of Balochistan. Roul maintains:

> Wilayat-e-Khorasan, the Islamic State (IS) affiliate in the borderlands of Afghanistan and Pakistan, is one of the terrorist group's strongest franchises. Bolstered by defections from the Taliban and boosted further in recent

months by an influx of foreign fighters fleeing defeat in Iraq and Syria, IS Khorasan Province (ISKP) is growing in strength and influence.

(2018, p. 3)

According to an estimate, there are between 2000 to 3000 ISKP operatives in Pakistan (Giustozzi, 2016). Though the security establishment is in denial about the presence of Islamic State in Pakistan, the footprints of ISKP, through a well organised regional collaboration/alliance with insurgent groups and sectarian extremist organisations in Balochistan, clearly show the emerging threat. The emerging threat of ISKP requires a precise assessment in terms of the dangers it currently poses and may unleash in future to the CPEC in Balochistan.

Terror Nexus of ISKP in Balochistan

In 2015, ISKP formally started to establish its footstep in Pakistan. There were several confirmed reports by the CIA and ISI that many insurgents and terrorist organisations were already in negotiation with the ISKP about entering into a strategic partnership by early 2015 (Intelligence Security Solution (ISS), 2015, p. 15). Mostly these organisations, willing to establish a partnership with ISKP, were based in the south-western province of Balochistan. The Islamic State preferred Balochistan to be its base and hub of operational activities instead of the "traditional jihadist safe heaven" of FATA which is the border region of Pakistan and Afghanistan in the KPK province (Intelligence Security Solution Report, 2015, p. 15).

ISKP's Alliances/Collaboration and the Demography of Balochistan

The main philosophy and motive behind the ISKP's alliances both with the Baloch insurgent organisations and the sectarian organisations is getting access for conducting terrorist activities and recruitment throughout the province. It is essential to highlight that broadly the demography of Balochistan is as such that the south, western, and eastern districts of Balochistan province are dominated by the Baloch ethnicity and north and north-western districts are dominated by Pashtuns (see Appendix 4). The hub of secessionist activity in Balochistan is mostly the Baloch dominated districts. In contrast, the sectarian organisations, like LeJ, enjoy a large population support base in Pashtun-dominated areas as traditionally Pashtuns are more inclined towards religious practices compared to Baloch, but of late, a substantial support base for LeJ is also available in the Baloch-dominated districts. Moreover, the number of well-established and traditional religious seminaries (madrassas), which serve as the recruiting bases of the terrorist organisations, are more in the Pashtun dominated districts of Balochistan. Another important aspect is that many former Afghan Taliban fighters of Pakistani heritage and TTP operatives also live in the Pashtun dominated districts because both these organisations had a high cadre of Pashtun ethnicity. These factors also help ISKP in getting well trained human resources from these areas. By entering into collaborations both with the secessionist Baloch insurgent organisation and sectarian organisations, the ISKP can dominate and operate throughout Balochistan's landscape.

Converging Interests/Ideologies in Balochistan – Anti-Pakistani State and Anti-Chinese Sentiment

The common ideology of the various militant organisations in Balochistan is based on being anti-Pakistani and anti-Chinese. The aim of insurgent Baloch organisations like the BLA is to destabilise the writ of the State in Balochistan so that the province may gain independence from Pakistan. After the signing of the MoU of CPEC in 2013, the insurgent organisations have shifted their focus to derail the CPEC in Balochistan by sabotaging CPEC related infrastructure, killing labourers working on the project and attacking the Chinese nationals in the province. In this way, owing to CPEC in Balochistan, the insurgent organisations have added an anti-Chinese dimension to their ideology as well. These anti-Chinese and anti-Pakistani motives are the phenomena which help the Baloch Insurgent organisation to affiliate with the ISKP. There were many attacks which have been jointly claimed by the ISKP and BLA[5]; for example, the one in November 2017 that targeted senior police officers in Balochistan (Zaheerul, 2017). Moreover, the nexus of BLA and ISKP have also claimed attacks such as the attack on the Christian church in Quetta in 2017 that killed 9 and injured about 57 people (Zaheerul, 2017).

The attacks by BLA in Balochistan, jointly conducted with ISKP, are still in progress with higher frequency, ferocity, and lethality. For example, 10 military personnel were killed in an attack led by BLA on 26 January 2022 on a security check post in the Kech district, roughly 600 km south of the provincial capital Quetta (Al Jazeera, 2022). Even more lethal attacks by the BLA were carried out on 2 February 2022, simultaneously targeting a Frontier Constabulary Fort/Camp (paramilitary force) in Panjgur town, located about 450 km south of Quetta, and another one in Noshki, about 330 km north of the first place of attack (Hashim, 2022). The Noshki assault was quashed on the next day, but the attack in Panjgur was only quelled after two days by the army. The army killed 13 terrorists, while 7 soldiers were also killed during these operations. Sheikh Rashid Ahmad, Pakistan's then federal interior minister, confirmed to the media that the BLA and the ISKP jointly carried out the attacks (The News, 2022).

Similarly, in an ISKP orchestrated suicide attack in Sibi on 8 March 2022, seven security personnel were killed, and 25 were wounded just minutes after the President of Pakistan, Dr Arif Alvi, attended a cultural event in town. According to a police official, the prime target of the attack was the president, and the possibility of the BLA being involved alongside the ISKP could not be ruled out (The Express Tribune, 2022; Durrani 2022, Interview, 11 August). In the same city, Sibi, a week later, on 15 March 2022, 4 Pakistani soldiers were killed, and 10 others were wounded due to the detonation of an improvised explosive device planted by the BLA (Majeed Brigade) near a convoy of security forces (Radio Free Europe/Radio Liberty, 2022).

ISKP has a much bigger goal for Pakistan to be part of Khorasan province of its global Islamic caliphate (Aamir, 2018; Adnan 2019, Interview, 14 January). ISKP carries an anti-Chinese agenda and continues to pose threats to the Chinese interests in Pakistan because it believes that China has "forcibly seized Muslims' rights" (Ma, 2015) in Xinjiang by not allowing the Uyghurs to practice the Islamic rituals and subjecting them to human rights violations. Moreover, China is an athe-ist country (Campbell, 2016) which is another reason, as per the ISKP ideology, to overrun the country (or at least Xinjiang province) by terrorist activities and make it part of the Khorasan province of the Islamic Caliphate (Aamir, 2018). This ide-ology allows the ISKP to motivate its rank and file to attack the Chinese interests in Balochistan.

The sectarian militant organisations involved in the killings in Balochistan sup-port the ISKP ideology, with varying strength, of establishing the Islamic Law in Pakistan as per their religious beliefs and ideology. They consider the Shiite and Zikri community in particular and other minorities in general, as non-believers of true Sharia and supporters of Western propaganda/influence against the Muslims.

There were confirmed reports of a meeting of the LeJ Al-Almi leadership with the ISIS leadership in Saudi Arabia (The Express Tribune, 2014) and Khost in Afghanistan (Jadoon, 2018, p. 41). LeJ claimed responsibility for many suicide attacks carried out in collaboration with ISKP; for example, in October 2016, a deadly suicide attack on a police training centre in Quetta which claimed the lives of 60 personnel and injured over 120 others. One of the officers in the security establishment of Pakistan confirmed that as per his reports, LeJ and ISKP work in collaboration in Balochistan (Durrani 2022, Interview, 11 August). The conver-gence of ideology and targets facilitate this collaboration (Zahid, 2017, pp. 6–7). "There are strong indications of the two groups possessing logistical and opera-tional synergy, which may [further] endure over time" (Jadoon, 2018, p. 41). For instance, the Pakistani Military claimed the killing of Salman Badeni, the alleged commander (regional chief) of LeJ in Balochistan, in May 2018, in the suburbs of the provincial capital Quetta. Badeni was involved in killing over 100 Hazaras and LEAs personnel (Yousafzai, 2018a). In response to the above-mentioned claim of ISPR, the LeJ spokesman confirmed the killing but stated that Badeni was not a member of LeJ, but instead a member of the ISKP (Yousafzai, 2018a) This shows a close operational collaboration between LeJ and ISKP in Balochistan. The other notable anti-Shiite militant groups cooperating with ISKP in Balochistan are the Saifullah Kurd faction of LeJ and Jaish-ul-Islam (Abdul Basit, 2018).

Apart from the insurgent and sectarian organisations, ISKP is also in collabora-tion with two of the splinter groups of TTP, namely Jundullah (Soldiers of God) (Abdul Basit, 2017, p. 24) and Jamat-ul-Ahrar (JuA) [Assembly of the Free] in the province of Balochistan (Jadoon, 2018, p. 43). In November 2014, the leadership of Jundullah met the Islamic State delegation in Balochistan (Abdul Basit, 2017, p. 24) and just six days after this meeting the former announced its pledge to the latter (The Express Tribune, 2014). Moreover, the ISKP tried to merge Pakistani Jundullah with the Iranian Jundullah during 2015/16 to further their aim of taking jihadist insurgency inside Iran (Giustozzi, 2018, p. 159). Like Jundullah, ISKP is also believed to have operational collaboration with JuA in Balochistan. For example, JuA killed four senior Police Officers of the Balochistan Police depart-ment on 13th July 2017 in Quetta, but it was also claimed by the ISKP on its Amaq News Agency Website (Yousafzai, 2017b). JuA was banned by the Government of Pakistan in 2016, and later the next year, the UN put JuA on its Al-Qaeda and ISIL sanctions list (Jadoon, 2018, p. 42). Both Jundullah and JuA's aim and ideology of disrupting the writ of the Government of Pakistan through Jihadist insurgency and creation of the Islamic Caliphate are suggestive of their collaboration with ISKP in Balochistan.

Balochistan as the Epicentre of ISKP's Activities

The province of Balochistan, because of the prevailing insurgency, the deteriorating law and order situation, sharing a porous border with Afghanistan and Iran, the presence of large numbers of Shiite and Zikri communities and above all CPEC (particularly Gwadar Port) provides the troika of ISKP, sectarian militant organisations, and insurgent organisations with a good opportunity of collaboration in conducting terrorist activities across the length and breadth of the province. Balochistan appears to be the epicentre of ISKP activities in Pakistan since the formal announcement, via an audio message, of its formation in 2015. The Af-Pak border area, a multi-faceted militant landscape, was an ideal breeding ground for the ISKP (Abdul Basit, 2017, p. 24). The ISKP preferred the south-western province of Balochistan as its base in Pakistan as opposed to FATA which earlier had been a safe haven for the Taliban, TTP, and Al Qaeda (Intelligence Security Solution Report, 2015, p. 15). The main reason for that was the focus of ongoing Operation Zarb-e-Azab by the Pakistani Military in FATA against the militants. Secondly, the easy alliance formation and collaboration owing to the convergence of aim and ideology with the already existing militant groups in the insurgency infested Balochistan was an added bonus for the ISKP. Balochistan is not only the base for ISKP in Pakistan but also the hub of terrorist activities and recruitment owing to its unique combination of the recent history of sectarian feuds, cultural divide, and sub-nationalists tendencies/clashes amongst the populace and the ongoing fifth round of insurgency. Schulz (2022) maintains that targeted attacks by ISKP on China-funded projects in Balochistan are already a great concern for Pakistani authorities.

India – An External Threat to CPEC

The external threats to CPEC are with regards to patronising/funding of the Baloch insurgents and splinter groups of TTP by India. The external support for these organisations helps them to grow faster (Ahsan 2019, Interview, 16 January). Pakistan asserts that India is a security threat to Balochistan in general, and CPEC projects including Gwadar Port, in particular (MOFA, 2022). It claims that India had been fomenting the insurgency in Balochistan, by providing funding and training to the militants in the camps based in Afghanistan[6], which was focused on sabotaging CPEC after 2013 (Iqbal, 2018, p. 208). This stance of Islamabad about security concerns for Balochistan/CPEC against the backdrop of alleged Indian involvement is well projected by the statements of the military and civilian leadership of the country alike (Ibrar et al., 2016). For instance, General Zubair Mehmood Hayat, Chairman Joint Chiefs of Staff Committee (CJCSC) of

Pakistan, stated, "RAW has established a special cell at a cost of $500 million to sabotage the CPEC and India is stoking chaos and anarchy in the region" (The Express Tribune, 2017).

Islamabad's claims of Indian involvement in sabotaging the CPEC are premised on several reasons (Kazmi 2018, Interview, 19 December). Foremost is that Indian leadership openly opposed CPEC as they consider it a challenge to Indian sovereignty. CPEC passes through Gilgit-Baltistan in Pakistan over which India lays claim for part of its Kashmir territory (Pandit, 2018). The Indian interest of influencing her sphere of interest in the region and gaining access to the Central Asian Energy corridor is of vital importance to the security of CPEC (Ahsan 2019, Interview, 16 January). In a bilateral meeting with Chinese President Xi Jinping, Indian PM Modi expressed his grave concerns over the CPEC project and was reported to have put across his opinions to Xi that both countries need to be sensitive about their strategic projects (Roy Chowdhury, 2017).

Moreover, the Indian Government and its advisors openly support the cause of an independent Balochistan. For example, during the Independence Day speech in 2016, Indian PM Narendra Modi highlighted the atrocities in Balochistan. Modi reiterated that his government would continue to highlight the Balochistan issue (Suhasini, 2016). Within a few hours, Modi's remarks were responded to by the Government of Pakistan and the then Pakistani Foreign Affairs Advisor Sartaj Aziz, who said that "Modi's reference to Balochistan, which is an integral part of Pakistan, only proves Pakistan's contention that India through intelligence agency RAW has been fomenting terrorism in Balochistan" (Suhasini, 2016). In the same context, the threatening remarks of Ajit Doval in 2014, the national security advisor to PM Modi, that if Pakistan carries out one Mumbai attack, it may lose Balochistan (Scroll.In, 2016) supports the idea that RAW has been active in Balochistan.

Pakistani LEAs captured an Indian spy, Kulbhushan Jadhav, from Balochistan in March 2016 and presented him to the world. Pakistan claimed that Jadhav, a serving naval officer and on deputation to the RAW, entered Balochistan through the Iranian border. They claimed that in Iran, he was based at Chabahar and working with proxies in Balochistan to sabotage CPEC, especially around Gwadar Port (Iqbal, 2018, pp. 207–208). The Indian government accepted that Jadhav was a former Navy officer but denied any involvement. Pakistan has questioned as to how Indian Naval Commander Kulbhushan Jadhav entered Pakistan without a visa on authentic Indian passport with a fake alias Hussain Mubarak Patel (TRT World, 2020; Ministry of Foreign Affairs (MOFA), 2020). The same valid Indian passport with a false Muslim identity[7] of Kulbhushan Jadhav was presented to the International Court of Justice by Pakistan as evidence of Jadhav's involvement in the terrorist activities in Balochistan (Kattan, 2020, p. 285). Pakistani claims of Indian involvement in sabotaging Gwadar Port are also supported by the growing Indian concern about CPEC particularly about a "potential Naval base at Gwadar to ensure Chinese maritime hegemony in the Indian Ocean" (Iqbal, 2018, p. 208).

The acceptance of the Indian Government that Jadhav had been in the Indian Navy supports the idea that he was not on a business trip to Balochistan as claimed by some of the Indian media. The death sentence to Jadhav, awarded by the Pakistani Military Court in April 2017 on the charges of spying, was challenged by India in the International Court of Justice (ICJ) on 8 May 2017. On 7 July 2019, the ICJ announced its decision after several detailed hearings. The court decided that Pakistan was under an obligation to provide, by means of its own choosing, effective review and reconsideration of the conviction and sentence of Mr. Jadhav (International Court of Justice (ICJ)(b), 2019). The case of Mr. Jadhav is still under legal review, in the light of ICJ's decision, in Pakistan (Ali, 2023).

Saudi Arabia and Gulf States

CPEC is confronted with another external threat from Saudi Arabia and the Gulf States, who have a history of financing prohibited sectarian groups in Pakistan, such as ASWJ and LeJ. They do so particularly in Balochistan, with the aim of countering Iranian influence in the province. This threat is no less than the threat posed by India to CPEC, but the Government of Pakistan has never highlighted it officially. There are numerous reasons for Pakistan's reticence to blame Saudi Arabia or any of the Gulf States. These include diplomatic ties and defence cooperation, financial aid packages received from these countries, business interests of the Pakistani political elite in these states, lucrative post-retirement settlement opportunities for the military's top brass, and political exile options in the Gulf region for the country's political establishment. Keeping these affiliations in view, now in Pakistan, not even journalists or bloggers dare to comment against any policy of Saudi Arabia or any of the Gulf States, especially the hardline stance against Shiite Muslims and their funding of banned sectarian organisations in Pakistan. For instance, the journalists who displayed pictures of Jamal Khashoggi on their social media profiles to highlight his murder during the visit of the Saudi Crown Prince Muhammad bin Salman (MBS) to Islamabad in February 2019, were investigated by the intelligence agencies for showing so-called disrespect to the "honourable" guest (Farmer, 2019a).

In Balochistan, Saudi Arabia, Qatar, and UAE are funding the religious seminaries (madrassas) run by militant organisations like LeJ, and ASWJ possessing strong anti-Shiite sentiments and followers of the Deobandi school of thought which is very close to Salafi school of thought[8] (Dorsey, 2019; Siddiqa, 2018, pp. 148–151; Zaka 2019, Interview, 03 January; Durrani 2022, Interview, 11 August). A majority of the children in Balochistan go to madrassas rather than the schools run by the government (Dorsey, 2019, p. 148). These madrassas are in much better condition than the government schools with most of them equipped with boarding facilities for the students. That is how the funding from Saudi Arabia and the Gulf States such as Qatar and UAE help the sectarian militants to dominate the educational landscape of Balochistan which ultimately results in the growth of these organisations, in numbers, by direct recruitment from these madrassas. Thus, they establish a strong foothold in Balochistan. One of the co-founders of ASWJ mentioned that the madrassas in Balochistan direct the Saudi funding to the militants. He further highlighted that the chief of LeJ in Balochistan, Molana Ramzan Mengal, notorious for killing Hazaras, also receives direct funding from the Saudis. Mengal moves in the protection of the police, and the police also guard his madrassa as the government/intelligence agencies consider a significant threat to his life from Iran (Dorsey, 2019, p. 148).

Moreover, the Saudi and UAE funding and backing to Jundullah in Balochistan helped it to further its anti-Iran agenda (Dorsey, 2019). Jundullah carries out attacks inside Iran, mostly targeting the Revolutionary Guards deployed for border protection. For example, in February 2019, 27 members of Iran's Revolutionary Guard were killed and injured another 13 of them in a suicidal attack perpetrated by Jaish al-Adl an offshoot of the Salafist militant group Jundullah. This incident took place in the province of Sistan and Baluchistan near Iran's border with Balochistan (Pakistan) (Eqbali & Rasmussen, 2019). Iran retaliates by firing rockets inside Pakistani territory (Balochistan) sometimes killing people living in the border areas (Dorsey, 2019; Yousafzai, 2017a, p. 154). At times this leads to firefights between the Pakistani Frontier Corps, deployed at the border, and the Iranian Revolutionary Guards. In this way, the security situation in Balochistan gets worse, and the militant organisations take full advantage of such situations by killing Hazaras and conducting attacks against Chinese or CPEC infrastructures. In a nutshell, "Saudi backing for the militants in Balochistan posed a problem for China ...... [by] taking the kingdom's proxy war with Iran to Balochistan threatened aggravating the already troubled security situation in CPEC's crown jewel" (Dorsey, 2019, p. 151).

Geography of Resistance and CPEC Alignment

The geography of resistance in Balochistan was transformed in 2006 after the killing of Nawab Akbar Bugti in the mountains of Kohlu district (Balochistan) by the Pakistani Military. The killing of Bugti reinforced the sense of nationalism amongst the Baloch, transcended tribal identities and incited the fifth round of insurgency in the province. Baloch tribes forgot about their centuries-old blood feuds and stood united against the Government of Pakistan. The grand Jirga (traditional assembly) convened by the Khan of Kalat in the aftermath of Bugti's killing in 2006 was attended by 84 Sardars (tribal chiefs) of Balochistan along with 300 influential personalities from the province (Shahid, 2006). This grand assembly condemned the killing in the strongest terms. This was the first-ever grand assembly of the tribal chiefs in Balochistan in the last 150 years. This showed a great leap towards Baloch nationalism.

The Baloch resistance/armed struggle which had been limited to the tribes and home districts of Marri, Mengal and Bugti in earlier rounds of insurgency spread all across the province. The transformation from tribalism to nationalism attracted growing participation in the armed struggle from the non-tribal districts like Gwadar, Kech, Panjgur, and Lasbela (Ahmad, 2012). For example, BLA which earlier had only Marri tribesmen as its rank and file now turned into a plural outfit composed of Marri, Bugti, Mangel, and Baloch from southern districts like Gwadar and the educated Baloch middle class from all over the province (Wani, 2016). The same happened to Allah Nazar's BLF and Brahamdagh's BRA. Over time these Baloch insurgent organisations gained popular support amongst the Baloch populated districts in southern, eastern, and western parts of Balochistan. At present, their active guerrilla fighters are spread all across Balochistan in small groups, and they maintain sleeper cells as well (Adeel 2019, Interview, 08 January; Durrani 2022, Interview, 11 August).

Likewise, the sectarian organisations which were earlier limited to Quetta city, north and north-western parts of Balochistan started spreading their influence in other parts of the province as a result of the increased inflow of funds from Saudi Arabia and other Gulf countries like UAE. Qatar and Oman. The number of madrassas rose exponentially in the province, including the Baloch dominated districts. At present, there are many Baloch leaders in the cadre of LeJ, ASWJ, and JuA. The collaborations of these insurgent and the sectarian organisations with each other and with ISKP has made the entire province (all 37 districts) accessible for the joint armed struggle focused after 2013 on sabotaging the CPEC and Chinese interests in Balochistan.

Gwadar and Makran Coastal Highway (Eastern Alignment of CPEC)

The eastern alignment of the corridor originates at Gwadar Port, passes through the Makran coastal highway towards Karachi and after passing through interior Sindh, southern, central and northern Punjab, Kashmir and Gilgit Baltistan it connects to China via the Khanjarab pass up in the north (see Appendix 5). On this eastern route, the Makran coastal highway (653 kms in length) linking Gwadar with Karachi is part of CPEC's eastern route in Balochistan. The security threats to the CPEC construction, workforce, and trade traffic on the Makran coastal highway had originated from the districts of Gwadar, Kech, Awaran, and Lasbela as these have immediate proximity to the CPEC eastern route alignment. Apart from these four districts, the two northern adjoining districts of Panjgur and Khuzdar also pose a grave security threat to the Makran coastal highway. These districts are very volatile as around 35 per cent of the attacks reported in Balochistan between 2007–2018 were carried out in these six southern districts (Zaka 2019, Interview, 03 January). The targets of these attacks were the Gwadar Port and its allied facilities, LEAs personnel, civilians, Chinese workers on Gwadar Port, railway tracks and government installations. The Baloch insurgent organisation like BLA, BRA, and BLF have a high presence in these districts, and they carry out regular attacks.

The sectarian militant organisations like LeJ, ASWJ, and Taliban splinter groups like the JuA are also increasing their presence in Khuzdar district which lies north of the Makran coastal belt. The growing nexus between the Baloch insurgent organisations and the sectarian militant organisation under the banner of ISKP may pose a more significant security risk to these areas in future.

Gwadar

Gwadar district and coastline are attractive areas to attack the CPEC infrastructure for the insurgent and sectarian organisations independently, and in collaboration with ISKP, because of the obvious geostrategic and economic significance of the

Gwadar Port for China and Pakistan. The insurgents have carried out many attacks on Gwadar in the past, but due to the high presence of security forces in the area, they could not inflict any enormous damage to Gwadar Port and allied facilities. ISKP has established itself in the Makran Division so that it can focus its terrorist activities onto the Gwadar Port (Aamir, 2018). This clearly shows the intentions of ISKP to disrupt and derail the CPEC by targeting Gwadar port in future.

Western Route Alignment of CPEC

The Western route of CPEC originates from Gwadar, runs through some southern districts (Turbat, Khuzdar, Kalat, Quetta, Zhob) and enters KPK province to link up with China in the north (see Appendix 5). The length of this route in Balochistan is over 1300 km. Towards the southern end of the route in the districts of Gwadar, Kalat, and Turbat the insurgent organisations are very active. In Quetta city, the capital of Balochistan, the sectarian organisations are involved in killing Hazaras and even targeting Chinese nationals. They have a strong foothold in Khuzdar and the neighbouring district of Mastung as well. The Taliban splinter group JuA is also based in Zhob district. Over the last few years, the terrorist attacks on CPEC infrastructure, suicide attacks on Chinese engineers, killing the labourers working on the road construction, targeted killings of Shiites and LEAs personnel were mainly focused along the western route of the CPEC.

Conclusion

The complex security situation in Balochistan directly threatens CPEC, the flagship project of BRI. There is a dire need to re-assess the security situation in Balochistan, by drawing some pertinent conclusions, and then devise a comprehensive strategy at the highest level in the country, maybe with Chinese collaboration, to deal with it in the context of CPEC. The attacks by the insurgent organisations like the BLA, BRA, and BLF in recent years have adversely impacted the Chinese investment projects in Balochistan. These attacks have also resulted in inhibiting the free movement of Chinese people in the region.

The dimension of the collaboration of these insurgent organisations amongst each other, with various sectarian organisations, splinter groups of TTP, and finally ISKP has added lethality to these attacks. The trend of suicide attacks, the common practice of Islamic State, in Balochistan during the recent years is the outcome of this collaboration. For example, a suicide attack claimed 149 dead and over 250 injured during an election campaign rally in Mastung town (Balochistan) in July 2018 (BBC News, 2018a). ISKP and LeJ claimed the responsibility of this second-deadliest terror attack in the history of the country.

The suicide attacks by ISKP will continue in Balochistan as apart from being a hub of militants and offering lucrative targets like ethnic Shite minorities, the province is the nucleus of Chinese investment for infrastructure development in Pakistan. ISKP will continue to further strengthen the collaboration with the local insurgent groups and the sectarian organisations and factions of TTP in

Balochistan. This collaboration poses a great security threat to CPEC. This strategy helps ISKP to have access to carry out attacks all across the length and breadth of Balochistan. Moreover, it attracts local recruitment for the organisation. These alliances are about to transform into enduring networks, considerably more potent, with the ability to move resources, human resources and logistics across the borders of Pakistan, Afghanistan, and Iran which may extend up to Xinjiang later if not checked effectively and in time.

The existing volatile security environment of Balochistan because of the presence of diverse actors, directly challenges the SCW doctrine of the Pakistani Army, especially after 2013, on which the army based its operations in Balochistan. The emerging puzzle is whether the Pakistani Army is conceptually equipped to fight these diverse threats to CPEC in Balochistan or not. If yes, what is the efficacy of the Pakistani Army's COIN strategies in the province? The next chapter(s) will unlock this puzzle that emerged from the rich description and assessment of the existing threats and their projected trajectory in the current chapter.

Notes

1 Dr Kaiser Bengali is a renowned economist in Pakistan and the former economic advisor to the Government of Balochistan. He also represented the province several times (as a non-statutory member) in the National Finance Commission (NFC).

2 Recognising the severe implications of porous borders, the Pakistani Military initiated the ambitious border security project, the chain-link fencing along the Pakistan-Afghanistan (Af-Pak) border, in 2017 as part of Operation Radd-ul-Fasaad. The project, which involves the construction of outposts and implementing a sophisticated surveillance system, initially valued at $483 million and later enhanced to $532 million (Ahmad Khan, 2021), was aimed at improving border security and preventing illegal cross-border movement (Yusufzai et al., 2018; ISPR, 2017c; Mir & Watkins, 2022). However, factors such as the strong retaliation of the Ghani and later Taliban government, border clashes with the Afghan security forces/Taliban, and the severe existing economic crisis in Pakistan have prevented the complete fencing of the Af-Pak border until now. The construction of a fence along the Pakistan-Iran border has not yet commenced.

3 In May 2019, the Islamic State fragmented the Islamic State Khorasan Province (ISKP) into separate branches for Pakistan, Afghanistan, and India, resulting in the creation of the Islamic State of Pakistan (ISPP) (Webber & Valle, 2022). In July 2021, ISPP Wali Abu Mahmood announced through a statement that the Islamic State (IS) had transferred Khyber Pakhtunkhwa (KPK) Province in Pakistan to the ISKP administration, prompting the group to claim responsibility for any future attacks in the area under the name of ISKP (Webber & Valle, 2022). However, Durrani (2022, Interview, 11 August), a police officer serving in the Counter Terrorism Department (CTD) of Balochistan Police, claims that there is still ambiguity regarding the Islamic State's brand name in Balochistan province. While sometimes the organization claims responsibility for terrorist attacks in Balochistan as ISPP, other times it is claimed as ISKP. Therefore, for clarity, the Islamic State in Balochistan will be referred to as ISKP throughout this book.

4 The terms Salafi and Wahhabi are often used interchangeably as Wahhabism is Salafism. (Ali & Sudiman, 2016).

5 Balochistan Liberation Army (BLA) is sometimes also known as Baluchistan National Army (BNA) and Balochistan National Liberation Army (Terrorism Research and Analysis Consortium (TRAC), 2019)

6 The training camps were reportedly established and operated during the era of the successive US-backed governments in Kabul before the US withdrawal from the country in 2021.
7 The forensic examination and evaluation of Kulbhushan Jadhav Indian passport with a fake alias was done by an independent expert named David Westgate. Mr Westgate had served as part of the United Kingdom Home Office and Immigration Intelligence Directorate for more than twenty-seven years, including serving on attachment to New Delhi and Karachi (Kattan, 2020, p. 285, footnote 10). Westgate's report established that the passport was authentic and issued by the government of India.
   Westgate concluded;

> Based upon my observations and my experience and examination of the passport, I am satisfied that it is an authentic Indian passport which I believe must have emanated from the Indian authorities.
>
> (2019, cited in ICJ(a), 2019, p. 28)

   The issuance of a valid Indian passport with a false Muslim identity for the express purpose of engaging in terrorist activities for the insurgents in Balochistan was an "abuse of process".
8 As Chris Blackburn (2006) notes, "the followers of the Deobandi sect of South Asia and the Wahhabi sect [Salafi] ……… follow an ultra-orthodox interpretation of the Islamic faith, which has often led them on a collision path with other sects …. " Although there are a few differences in jurisprudence between Deobandi and Salafi however, the former aligns with the global Salafi and Wahhabi movements and is hence rightly known as the "South Asian variant of Salafis/Wahhabis (Syed, 2016)".

References

Aamir, A. (2018). 'ISIS Threatens China-Pakistan Economic Corridor', *China-US Focus*, 17 August. Available at: https://www.chinausfocus.com/peace-security/isis-threatens-china-pakistan-economic-corridor (Accessed: 16 April 2022).

Abdul Basit (2016). 'Pakistan', in Gunaratna, R. & Stefanie, K. L. (eds.) *Handbook of Terrorism in the Asia–Pacific*. London: Imperial College Press, pp. 215–242.

Abdul Basit (2017). 'IS Penetration in Afghanistan-Pakistan Assessment, Impact and Implications'. *Perspectives on Terrorism*, 11 (3), pp.19–39. Available at: https://www.jstor.org/stable/26297839?seq=1#metadata_info_tab_contents (Accessed: 25 July 2022).

Abdul Basit (2018). 'Threat to Pakistan's Internal Security by The Islamic State of Khorasan'. *Pakistan Politico,*7 August. Available at: http://pakistanpolitico.com/threat-to-pakistans-internal-security-by-the-islamic-state-of-khorasan/ (Accessed: 21 April 2022).

Adeney, K. (2012). 'A Step Towards Inclusive Federalism in Pakistan? The Politics of the 18th Amendment', *Publius. The Journal of Federalism*, 42 (4), (Fall 2012), pp. 539–565. doi: 10.1093/publius/pjr055.

Ahmad, M. (2012). 'Balochistan: Middle-class rebellion', *Dawn,* 05 June. Available at: https://www.dawn.com/news/723987/balochistan-middle-class-rebellion (Accessed: 19 April 2022).

Ahmad Khan, M. (2021). 'Achievements of Radd-ul-Fasaad', *The Nation*, 15 March. Available at: https://www.nation.com.pk/15-Mar-2021/achievements-of-radd-ul-fasaad (Accessed: 20 April 2022).

Aid to the Church in Need (ACN) (2022). *Balochistan is deadly for Christians in Pakistan.* Available at: https://www.churchinneed.org/balochistan-is-deadly-for-christians-in-pakistan/ (Accessed: 20 April 2023).

https://dailytimes.com.pk/970709/a-biggerthreat/amp/ (Accessed: 22 April 2023).

Ali, M. B. & Sudiman, M. S. (2016). 'Salafis and Wahhabis: Two Sides of the Same Coin?', *RSIS Commentary*, 254 (11 October). Available at: https://www.rsis.edu.sg/wp-content/uploads/2016/10/CO16254.pdf (Accessed: 27 May 2022).

AlJazeera (2022). *Several Pakistani soldiers killed in southwest checkpost attack*. Available at: https://www.aljazeera.com/news/2022/1/28/pakistan-soldiers-killed-balochistan-southwest-checkpost-attack (Accessed: 20 April 2022).

Arduino, A. (2017). 'China-Pakistan Economic Corridor: Security and Inclusive Development Needed', *Asia Dialogue*, 18 July. Available at: http://theasiadialogue.com/2017/07/18/china-pakistan-economic-corridor-security-and-inclusive-development-needed/ (Accessed: 27 April 2022).

Associated Press of Pakistan (APP) (2019). *Pakistan to fence border with Iran to evade Omara like attacks: Qureshi*. Available at: https://www.app.com.pk/pakistan-to-fence-border-with-iran-to-evade-omara-like-attacks-qureshi/ (Accessed: 20 April 2022).

Basit, S. H. (2019). 'Terrorizing the CPEC: Managing Transnational Militancy in China–Pakistan Relations', *The Pacific Review*, 32 (4), pp. 694–724. doi: 10.1080/09512748.2018.1516694.

BBC News (2003). *Pakistani group joins US terror list*. Available at: http://news.bbc.co.uk/1/hi/world/south_asia/2711239.stm (Accessed: 02 March 2022).

BBC News (2015). *Pakistan dam builders shot dead in Balochistan*. Available at: https://www.bbc.co.uk/news/world-asia-32263534 (Accessed: 02 July 2022).

BBC News (2018a). *Pakistan mourns 149 dead in country's second deadliest terror attack*. Available at: https://www.bbc.co.uk/news/world-asia-44847295 (Accessed: 12 March 2022).

BBC News (2018b) *Karachi attack: China consulate attack leaves four dead*. Available at: https://www.bbc.co.uk/news/world-asia-46313136 (Accessed: 27 March 2022).

BBC News (2022) *Pakistan attack: China condemns killing of tutors in Pakistan blast*. Available at: https://www.bbc.com/news/world-asia-61225678 (Accessed: 20 April 2023).

Blackburn, C. (2006). 'Terrorism in Bangladesh: The Region and Beyond', paper presented at the Policy Exchange Conference in London and also published in *New Age* (September 22, 2005).

Campbell, C. (2016) 'China's Leader Xi Jinping Reminds Party Members to Be 'Unyielding Marxist Atheists', *TIME*, 25 April. Available at: https://time.com/4306179/china-religion-freedom-xi-jinping-muslim-christian-xinjiang-buddhist-tibet/ (Accessed: 26 April 2022).

Chandran, N. (2018) 'China defends its alliance with Pakistan after consulate attack', *CNBC*, 26 November. Available at: https://www.cnbc.com/2018/11/26/china-defends-alliance-with-pakistan-after-karachi-consulate-attack.html (Accessed: 20 April 2022).

Chaudhry, S. (2018) 'Govt lifts ban on ASWJ, unfreezes assets of its chief Ahmed Ludhianvi', *The Express Tribune*, 27 June. Available at: https://tribune.com.pk/story/1744294/1-govt-lifts-ban-aswj-unfreezes-assets-chief-ahmed-ludhianvi/ (Accessed: 15 March 2019).

Chaziza, M. (2016). 'China–Pakistan Relationship: A Game-Changer for the Middle East?', *Contemporary Review of the Middle East*, 3 (2), pp. 147–161. doi: 10.1177/2347798916638209.

Dawn (2018) *Pak Army shall ensure security of CPEC at all costs, Gen Bajwa tells President Xi Jinping*. Available at: https://www.dawn.com/news/1433887 (Accessed: 20 April 2022).

Deutsche Welle (DW) (2015) *Brahamdagh Bugti: 'China-Pakistan deal usurps Balochistan's resources'*. Available at: https://www.dw.com/en/brahamdagh-bugti-china-pakistan-deal-usurps-balochistans-resources/a-18405846 (Accessed: 27 March 2022).

Dorsey, J. M. (2019). *China and the Middle East: Venturing into the Maelstrom*. Palgrave Macmillan.

Eqbali, A. & Rasmussen, S. E. (2019) 'Suicide Bombing Kills Revolutionary Guards in Restive Iranian Province', *The Wall Street Journal*, 13 February. Available at: https://www.wsj.com/articles/suicide-bombing-kills-revolutionary-guards-in-restive-iranian-province-11550088294 (Accessed: 21 September 2022).

Fair, C. C. (2014). *Fighting to the End: The Pakistan Army's Way of War*. USA: Oxford University Press.

Farmer, B. (2019a) 'Pakistan investigates journalists for online campaign to 'disrespect' Saudi Crown Prince with Khashoggi pictures', *The Telegraph*, 28 March. Available at: https://www.telegraph.co.uk/news/2019/03/28/pakistan-investigates-journalists-online-campaign-disrespect/ (Accessed: 19 April 2022).

Farmer, B. (2019b) 'Balochistan Liberation Army: Pakistan hails US for terrorist group designation', *The National,* 03 July. Available at: https://www.thenational.ae/world/asia/balochistan-liberation-army-pakistan-hails-us-for-terrorist-group-designation-1.882355 (Accessed: 17 July 2022).

Giustozzi, A. (2016) 'The Islamic State in "Khorasan": A nuanced view', *RUSI Commentary*, 05 February. Available at: https://rusi.org/commentary/islamic-state-khorasan-nuanced-view (Accessed: 29 March 2022).

Giustozzi, A. (2018). *The Islamic State in Khorasan: Afghanistan, Pakistan and the New Central Asian Jihad*. London: C Hurst & Co Publishers Ltd.

Grare, F. (2006) *Pakistan: The Resurgence of Baluch Nationalism*. Carnegie Endowment for International Peace. Available at: https://carnegieendowment.org/files/CP65.Grare.FINAL.pdf (Accessed: 22 July 2022).

Grare, F. (2013) *Balochistan: The State versus the Nation*, Carnegie Endowment for International Peace. Available at: https://carnegieendowment.org/2013/04/11/balochistan-state-versus-nation-pub-51488 (Accessed: 29 July 2022).

Gul, A. (2022) 'China Urges Pakistan to Ensure Security of Chinese Working on Bilateral Projects', *Voice of America (VOA)*, 02 November. Available at: https://www.voanews.com/a/china-urges-pakistan-to-ensure-security-of-chinese-working-on-bilateral-projects/6817008.html (Accessed: 8 December 2022).

Haider, M. (2016) 'Army aware of hostility against CPEC, will protect it at any cost: Gen Raheel', *Dawn,* 02 June. Available at: https://www.dawn.com/news/1262298 (Accessed: 27 April 2022).

Harrison, S. (1978). 'Nightmare in Baluchistan', *Foreign Policy*, 32 (Autumn), pp. 136–160.

Hasan, S. S. (2009) 'Pakistan "extremist" is shot dead', *BBC News,* 17 August. Available at: http://news.bbc.co.uk/1/hi/8205158.stm (Accessed: 19 June 2022).

Hasan, S. S. (2012) 'Pakistan bans Ahle Sunnah Wal Jamaat Islamist group', *BBC News,* 10 March. Available at: https://www.bbc.co.uk/news/world-asia-17322095 (Accessed: 15 June 2022).

Hashim, A. (2016) 'Pakistan: BLF chief Baloch says Indian help 'welcome'', *AlJazeera News,* 30 September. Available at: https://www.aljazeera.com/news/2016/09/pakistan-blf-chief-baloch-indian-160929153641899.html (Accessed: 15 August 2022).

Hashim, A. (2017) 'Bomb and gun attack on Quetta church kills eight', *AlJazeera,* 17 December. Available at: https://www.aljazeera.com/news/2017/12/pakistan-quetta-church-hit-suicide-attack-171217082230934.html (Accessed: 15 August 2022).

Hashim, A. (2018) 'Four killed in attack on Christians in Pakistan's Quetta', *AlJazeera,* 02 April. Available at: https://www.aljazeera.com/news/2018/04/killed-attack-christians-pakistan-quetta-180402161157315.html (Accessed: 15 August 2022).

Hashim, A. (2022) 'Pakistani forces battle gunmen after Balochistan checkpost raids', *Aljazeera*, 03 February. Available at: https://www.aljazeera.com/news/2022/2/3/pakistani-forces-battle-gunmen-after-balochistan-checkpost-raids (Accessed: 20 April 2022).

Iaccino, L. (2016) 'Baloch leader calls citizens "most oppressed in world", urges halt to aid to Pakistan', *International Business Times,* 08 February. Available at: https://www.ibtimes.co.uk/balochistan-baloch-leader-calls-citizens-most-oppressed-world-urges-halt-aid-pakistan-1508447 (Accessed: 27 April 2022).

Ibrar, M., Mi, J., Rafiq, M. & Karn, A. L. (2016) 'The China-Pakistan Economic Corridor: Security Challenges', *2nd Asia-Pacific Management and Engineering Conference (APME).* Shanghai China, 24–25 December. Pennsylvania, PA: DEStech Publications. ISBN: 978-1-60595-434-9. Available at: https://pdfs.semanticscholar.org/461c/f8f196a10ed90e06bd-2f563b637bc03b7d0b.pdf (Accessed: 21 April 2022).

Intelligence Security Solution (ISS) (2015) *Threat Assessment: Convergence of China's Economic Globalisation Strategy and Encroaching ISIS Aspirations for the Asian Caliphate.* Available at: http://issrisk.com/wp-content/uploads/2015/10/Oct2015-China-ISIS-Convergence-Executive-Summary.pdf (Accessed: 26 March 2022).

International Court of Justice (ICJ) (a) (2019) *Public sitting held on Tuesday 19 February 2019, at 10 a.m., at the Peace Palace, President Yusuf presiding, in the Jadhav case (India v. Pakistan) – VERBATIM RECORD.* Available at: https://www.icj-cij.org/files/case-related/168/168-20190219-ORA-01-00-BI.pdf (Accessed: 01 August 2022).

International Court of Justice (ICJ) (b) (2019) *Jadhav (India v. Pakistan): Overview of The Case.* Available at: https://www.icj-cij.org/en/case/168 (Accessed: 01 August 2022).

International Crisis Group (2018) *China-Pakistan Economic Corridor: Opportunities and Risks (Asia Report No. 297).* Available at: https://www.crisisgroup.org/asia/south-asia/pakistan/297-china-pakistan-economic-corridor-opportunities-and-risks (Accessed: 02 April 2022).

Iqbal, K. (2018). 'Securing CPEC: Challenges, Responses and Outcomes', in Arduino, A. & Gong, X. (eds.) *Securing the Belt and Road Initiative: Risk Assessment, Private Security and Special Insurances along the New Wave of Chinese Outbound Investments.* Singapore: Palgrave Macmillan, pp. 197–214.

Jadoon, A. (2018). *Allied & Lethal: Islamic State Khorasan's Network and Organizational Capacity in Afghanistan and Pakistan.* Combating Terrorism Center at West Point.

Jamal, U. (2022) 'Does Pakistan Have the Capability to Secure CPEC Projects?', *The Diplomat,* 22 July. Available at: https://thediplomat.com/2022/07/does-pakistan-have-the-capability-to-secure-cpec-projects/ (Accessed: 20 April 2023).

Janjua, H. & Shams, S. (2018) 'Why Baloch separatists are against China', *Deutsche Welle (DW),* 23 November. Available at: https://www.dw.com/en/china-consulate-attack-why-pakistans-baloch-separatists-are-against-beijing/a-46424112 (Accessed: 20 April 2023).

Kattan, V. (2020) 'Jadhav Case (India v. Pakistan)', *American Journal of International Law,* 114 (2), pp. 281–287. doi: https://doi.org/10.1017/ajil.2020.6. Published online by Cambridge University Press on 16 April 2020. Available at: https://www.cambridge.org/core/journals/american-journal-of-international-law/article/jadhav-case-india-v-pakistan/DD-5928FD5A298073FAC9970811419AED (Accessed: 02 August 2022).

Khaleej Times (2019) *Gunmen kill at least 14 bus passengers in Pakistan.* Available at: https://www.khaleejtimes.com/international/pakistan/gunmen-kill-at-least-14-bus-passengers-in-pakistan (Accessed: 11 June 2022).

Khan, A. (2009). 'Renewed Ethnonationalist Insurgency in Balochistan, Pakistan: The Militarized State and Continuing Economic Deprivation', *Asian Survey; Berkeley,* 49 (6), pp. 1071–1091. doi: 10.1525/as.2009.49.6.1071.

Khan Wajahat, S. (2018) 24 November. Available at: https://twitter.com/WajSKhan/status/1066307529905446914 (Accessed: 7 April 2022).

Kulkarni, S. (2022) 'The China-Pakistan Economic Corridor 2.0 – A New Risk Management Strategy', *China Global South Project*, 02 December. Available at: https://chinaglobalsouth.com/analysis/the-china-pakistan-economic-corridor-2-0-a-new-risk-management-strategy/ (Accessed: 20 April 2022).

Lieven, A. (2017). 'Counter-Insurgency in Pakistan: The Role of Legitimacy', *Small Wars & Insurgencies*, 28 (1), pp. 166–190. doi: 10.1080/09592318.2016.1266128.

Ma, A. (2015) 'ISIS Appears to Target Chinese Muslims in New Recruiting Effort', *Huffington Post*, 08 December. Available at: https://www.huffingtonpost.co.uk/entry/isis-recruiting-chinese-muslims_us_5666e238e4b079b2818ff5db?guce_referrer=aHR0cHM6Ly93d3-cuY2hpbmF1c2ZvY3VzLmNvbS9wZWFjZS1zZWN1cml0eS9pc2lzLXRocmocmVhdGVu-cy1jaGluYS1tdXNsaW1zLXdYWtpc3Rhbi1Y29ub21pYy1jb3JyaWRvci5odG1s&guce_referrer_sig=AQA-AAFY5ezJzM6WXBeIsIHA_0sU57qs2tSxykS4TqxDTQL1w992qhE_-HI2RjSfCRUU9bDKFOkEJA9WyZOG4caSyaDHgBGOVVEu04MXLTu9ByHhWQfn04eOknPAloF-vUe8Z0qH1B0540QTDVpBPsaPWsM2EjfEYdEr5xpnvinokMVdDu&_guc_consent_skip=1569922445 (Accessed: 11 March 2022).

Mashkoor, A. (2017). 'The Modern Thesis of Pakistan Politics'. *Munich Personal RePEc Archive (MPRA)*, Paper No. 80852. Available at: https://mpra.ub.uni-muenchen.de/80852/1/MPRA_paper_80852.pdf (Accessed: 21 July 2022).

Masood, S. (2019) 'Pakistan Market Bomb Blast Kills at Least 16 People in Quetta', *The New York Times*, 12 April. Available at: https://www.nytimes.com/2019/04/12/world/asia/pakistan-bombing-market.html?rref=collection%2Ftimestopic%2FLashkar-e-Jhangvi&action=click&contentCollection=timestopics&region=stream&module=stream_unit&version=latest&contentPlacement=1&pgtype=collection (Accessed: 27 April 2022).

McKirdy, E. & Saifi, S. (2019) 'At least 20 killed in market blast in Pakistani city of Quetta', *CNN*, 12 April. Available at: https://edition.cnn.com/2019/04/12/asia/quetta-market-blast-intl/index.html (Accessed: 15 June 2022).

Mir, A. & Watkins, A. (2022). 'Afghanistan-Pakistan Border Dispute Heats Up', *United States Institute of Peace (USIP)*, 12 January. Available at: https://www.usip.org/publications/2022/01/afghanistan-pakistan-border-dispute-heats (Accessed: 20 April 2023).

National Commission for Human (NCHR) (2018). *Understanding the agonies of ethnic Hazaras*. Available at: https://nchr.gov.pk/wp-content/uploads/2019/01/HAZARA-REPORT.pdf; https://hazaraunitedmovement.files.wordpress.com/2018/03/understanding-the-agonies-of-the-ethnic-hazaras-doc.pdf (Accessed: 22 August 2022).

News Intervention Bureau (2023) *We do not beg for justice from Pakistan: Dr. Allah Nazar Baloch*. Available at: https://www.newsintervention.com/we-do-not-beg-for-justice-from-pakistan-dr-allah-nazar-baloch/ (Accessed: 20 April 2023).

Pakistan Ministry of Foreign Affairs (MOFA). (2020). *Judgement of International Court of Justice on Commander Kulbhushan Jadhav*. Available at: http://mofa.gov.pk/judgement-of-international-court-of-justice-on-commander-kulbhushan-jadhav/ (Accessed: 05 August 2022).

Pakistan Ministry of Foreign Affairs (MOFA). (2022). *Pakistan categorically rejects the Indian MEA's preposterous comments on Pakistan-China Joint Statement of 6 February 2022*. Available at: https://mofa.gov.pk/pakistan-categorically-rejects-the-indian-meas-preposterous-comments-on-pakistan-china-joint-statement-of-6-february-2022/ (Accessed: 20 April 2023).

Pakistan Today (2017) *China's embassy in Pakistan warns of 'terror attacks' on its nationals*. Available at: https://archive.pakistantoday.com.pk/2017/12/08/chinas-embassy-in-pakistan-warns-of-terror-attacks-on-its-nationals/ (Accessed: 17 April 2022).

Pakistan. National Counter Terrorism Authority (NACTA) (2019). *73 Organizations Proscribed by Ministry of Interior u/s 11-B-(1) r/w Schedule-I, ATA 1997*. Available at: https://nacta.gov.pk/wp-content/uploads/2017/08/Proscribed-OrganizationsEng.pdf (Accessed: 17 April 2022).

Pandit, R. (2018) 'India expresses strong opposition to China Pakistan Economic Corridor, says challenges Indian sovereignty', *The Economic Times*, 12 July. Available at: https://economictimes.indiatimes.com/news/defence/india-expresses-strong-opposition-to-china-pakistan-economic-corridor-says-challenges-indian-sovereignty/articleshow/57664537.cms (Accessed: 13 April 2022).

Radio Free Europe Radio Liberty (RFERL) (2018) *Pakistan's Army Kills Commander of Islamist Militant Group in Balochistan*. Available at: https://www.rferl.org/a/pakistan-army-kills-commander-islamist-militant-group-lashkar-e-jhangvi-balochistan-badeni/29231434.html (Accessed: 17 July 2022).

Radio Free Europe Radio Liberty (RFERL) (2022) *Roadside Bomb Kills At Least Four Pakistani Soldiers, Wounds 10 In Southwestern Pakistan*. Available at: https://www.rferl.org/a/pakistan-balochistan-bomb-soldiers-killed/31754412.html (Accessed: 20 April 2022).

Roul, A. (2018) 'Islamic State Gains Ground in Afghanistan as Its Caliphate Crumbles Elsewhere'. *Terrorism Monitor; James Town Foundation*, 16 (2), (January 26), pp. 3–5. Available at: https://jamestown.org/wp-content/uploads/2018/01/Terrorism-Monitor-January-26-2018.pdf?x97873 (Accessed: 27 April 2022).

Roy Chowdhury, A. (2017) 'CPEC: The bumpy new trade route between China and Pakistan', *The Indian Express*, 16 May. Available at: https://indianexpress.com/article/research/cpec-the-bumpy-new-trade-route-between-china-and-pakistan-4373203/ (Accessed: 17 March 2022).

Saeed, A. (2022) 'A Bigger Threat', *Daily Times*, 23 July. Available at: https://dailytimes.com.pk/970709/a-biggerthreat/amp/ (Accessed: 22 April 2023).

Samad, Y. (2014). 'Understanding the Insurgency in Balochistan', *Commonwealth & Comparative Politics*, 52 (2), pp. 293–320. doi: 10.1080/14662043.2014.894280.

Samad, Y., Singh, Y. & Samad (2015). 'Understanding the Insurgency in Balochistan', in Long, R. D. & Talbot, I. (eds.) *State and Nation-Building in Pakistan: Beyond Islam and Security*. Abingdon, Oxon: Routledge, pp. 118–145.

Schulz, D. (2022) 'ISKP's Propaganda Threatens Asia's Security Apparatus', *Stimson*, 04 October 4. Available at: https://www.stimson.org/2022/iskps-propaganda-threatens-asias-security-apparatus/ (Accessed: 20 April 2022).

Scroll.In (2016) *Watch Ajit Doval say if Pakistan does 'one Mumbai' it may lose Balochistan*, Available at: https://scroll.in/video/801447/watch-ajit-doval-say-if-pakistan-does-one-mumbai-it-may-lose-balochistan (Accessed: 11 April 2022).

Shafqat Ali. (2023) 'Jadhav, the terror face of India', *The Nation*, 03 March. Available at: https://www.nation.com.pk/03-Mar-2023/jadhav-the-terror-face-of-india (Accessed: 20 April 2023).

Shah Murtaza, A. (2011) 'Hyrbyair Marri wins political asylum case', *The News International,* 30 January. Available at: https://www.thenews.com.pk/print/90818-hyrbyair-marri-wins-political-asylum-case (Accessed: 27 April 2022).

Shahid, S. (2006) 'Grand jirga in Kalat decides to move ICJ', *Dawn,* 22 September. Available at: https://www.dawn.com/news/211514 (Accessed: 19 March 2022).

Siddiqa, A. (2018). 'Pakistani Madrasas: Ideological Stronghold for Saudi Arabia and Gulf States' in C. Jaffrelot & L. Louer (ed.), *Pan-Islamic Connections: Transnational Networks between South Asia and the Gulf* (Online ed). Oxford Academic, pp. 49–72 https://doi.org/10.1093/oso/9780190862985.003.0003

Small, A. (2015). *The China-Pakistan Axis: Asia's New Geopolitics*. Oxford University Press.

South Asia Terrorism Portal (SATP) (2017a) *Lashkar-e-Jhangvi (LeJ)-Pakistan*. Available at: https://www.satp.org/terrorist-profile/pakistan/lashkar-e-jhangvi-lej (Accessed: 19 March 2022).

South Asia Terrorism Portal (SATP) (2017b) *Sipah-e- Sahaba Pakistan (SSP)-Pakistan*. Available at: https://www.satp.org/terrorist-profile/pakistan/sipah-e-sahaba-pakistan-ssp (Accessed: 21 March 2022).

Stanford Center for International Security and Cooperation (CISAC) (2018a) *Lashkar-e-Jhangvi (LeJ)*. Available at: https://cisac.fsi.stanford.edu/mappingmilitants/profiles/lashkar-e-jhangvi-lej#highlight_text_9533 (Accessed: 27 July 2022).

Stanford Center for International Security and Cooperation (CISAC) (2018b) *The Islamic State in the Khorasan Province (IS-KP)*. Available at: https://cisac.fsi.stanford.edu/mappingmilitants/profiles/islamic-state-khorasan-province#_edn1 (Accessed: 29 July 2022).

Stanford Center for International Security and Cooperation (CISAC) (2019) *Balochistan Liberation Army (BLA)*. Available at: https://cisac.fsi.stanford.edu/mappingmilitants/profiles/balochistan-liberation-army?highlight=Al+Nusrah+front#note53 (Accessed: 12 July 2022).

Stanford University (2015a) *Mapping Militant Organizations: Balochistan Republican Army (BRA)*. Available at: http://web.stanford.edu/group/mappingmilitants/cgi-bin/groups/view/571?highlight=baloch#cite26 (Accessed: 14 July 2022).

Stanford University (2015b) *Mapping Militant Organizations: Baluchistan Liberation Front (BLF)*. Available at: http://web.stanford.edu/group/mappingmilitants/cgi-bin/groups/view/457?highlight=baloch#note6 (Accessed: 17 July 2022).

Suhasini, H. (2016) 'In policy shift, Narendra Modi brings up Balochistan again', *The Hindu*, 16 August. Available at: https://www.thehindu.com/news/national/In-policy-shift-Narendra-Modi-brings-up-Balochistan-again/article14572650.ece (Accessed: 17 March 2022).

Syed Baqir, S. (2015) 'China vows support for CPEC security', *Dawn*, 13 November. Available at: https://www.dawn.com/news/1219331 (Accessed: 02 June 2022).

Syed Baqir, S. (2016) 'Civil-military differences hold up CPEC security plan', *Dawn*, 19 September. Available at: https://www.dawn.com/news/1284724 (Accessed: 03 June 2022).

Syed, J. (2016) 'Targeted Killings in Bangladesh: Diversity at Stake', *University of Huddersfield*, Available at: http://eprints.hud.ac.uk/id/eprint/28224/1/Syed%202016%20HuffPo%20Bangladesh.pdf (Accessed: 20 April 2023).

Terrorism Research and Analysis Consortium (TRAC) (2019) *Baluchistan National Army*. Available at: https://www.trackingterrorism.org/group/baluchistan-national-army (Accessed: 26 March 2022).

The Balochistan Post (2018a) *Attack on Chinese consulate natural reaction against oppression – Dr Allah Nazar*. Available at: http://thebalochistanpost.net/2018/11/attack-on-chinese-consulate-natural-reaction-against-oppression-dr-allah-nazar/ (Accessed: 02 March 2022).

The Balochistan Post (2018b) *Baloch armed organisations form umbrella organisation*. Available at: http://thebalochistanpost.net/2018/11/baloch-armed-organisations-form-umbrella-organisation/ (Accessed: 27 April 2022).

The Express Tribune (2014) *Jundullah vows allegiance to Islamic State*. Available at: https://tribune.com.pk/story/792872/jundullah-vows-allegiance-to-islamic-state/ (Accessed: 10 March 2022).

The Express Tribune (2017) *India spending $500 million to sabotage CPEC: Gen Zubair.* Available at: https://tribune.com.pk/story/1557930/1-indias-raw-aiming-undermine-cpec-cjcsc/ (Accessed: 17 July 2022).

The Express Tribune (2022) *Seven martyred in Sibi explosion.* Available at: https://tribune.com.pk/story/2346992/seven-martyred-in-sibi-explosion (Accessed: 20 April 2022).

The News (2022) *Alert issued after terror attacks in Balochistan.* Available at: https://www.thenews.com.pk/print/930580-alert-issued-after-attacks-in-balochistan (Accessed: 20 April 2022).

Tolo News (2018) *Baloch Separatist Leader Killed in Kandahar Attack: Reports.* Available at: https://tolonews.com/index.php/afghanistan/baloch-separatist-leader-killed-kandahar-attack-reports (Accessed: 11 April 2022).

TRT World (2020) *Pakistan invites India to file review against spy Jadhav's conviction.* Available at: https://www.trtworld.com/asia/pakistan-invites-india-to-file-review-against-spy-jadhav-s-conviction-37993 (Accessed: 01 August 2022).

UK. Home Office. (2021). *Proscribed Terrorist Organisations.* Available at: https://researchbriefings.files.parliament.uk/documents/SN00815/SN00815.pdf (Accessed: 28 April 2022).

United Nations (UN) Security Council. (2015). *Security Council Committee pursuant to resolutions 1267 (1999) 1989 (2011) and 2253 (2015) concerning ISIL (Da'esh) Al-Qaida and associated individuals groups undertakings and entities.* Available at: https://web.archive.org/web/20161018061034/https:/www.un.org/sc/suborg/en/sanctions/1267/aq_sanctions_list/summaries/entity/lashkar-i-jhangvi-(lj) (Accessed: 19 April 2022).

Wani, S. A. (2016). 'The Changing Dynamics of the Baloch Nationalist Movement in Pakistan: From Autonomy Toward Secession', *Asian Survey*, 56 (5), pp. 807–832. doi: 10.1525/as.2016.56.5.807.

Webber, L. & Valle, R. (2022) 'Islamic State Khorasan's Expanded Vision in South and Central Asia', *The Diplomat*, 26 August. Available at: https://thediplomat.com/2022/08/islamic-state-khorasans-expanded-vision-in-south-and-central-asia/ (Accessed: 20 April 2023).

Wu, W. (2018) 'Beijing condemns suicide attack on bus carrying Chinese engineers in Pakistan', *South China Morning Post,* 11 August. Available at: https://www.scmp.com/news/china/diplomacy-defence/article/2159291/bus-carrying-chinese-engineers-targeted-pakistan (Accessed: 21 April 2022).

Yousafzai, G. (2017a) 'Pakistan says Iranian mortar attack kills civilian', *Reuters,* 27 May. Available at: https://www.reuters.com/article/us-pakistan-iran-border/pakistan-says-iranian-mortar-attack-kills-civilian-idUSKBN18N0FM (Accessed: 01 March 2022).

Yousafzai, G. (2017b) 'Gunmen kill four police in Pakistani city of Quetta', *Reuters (India),* 13 July. Available at: https://in.reuters.com/article/pakistan-shooting-idINKBN19Y13V?feedType=RSS&feedName=southAsiaNews (Accessed: 11 March 2022).

Yousafzai, G. (2018a) 'Pakistani army kills senior militant, seven suicide bombers', *Reuters (UK),* 17 May. Available at: https://uk.reuters.com/article/uk-pakistan-militants/pakistani-army-kills-senior-militant-seven-suicide-bombers-idUKKCN1II0DH (Accessed: 10 April 2022).

Zaheerul, H. (2017) 'Quetta Church Attack Planned by Daesh and BNA', *Asian Tribune,* 18 December. Available at: http://www.asiantribune.com/node/91384 (Accessed: 11 March 2022).

Zahid, F. (2017) 'Lashkar-e-Jhangvi Al-Alami: A Pakistani Partner for Islamic State'. *Terrorism Monitor; James Town Foundation,* 15 (2), (January 27), pp. 6–7. Available at: https://jamestown.org/wp-content/uploads/2017/02/TM_January_27_2017.pdf?x97873 (Accessed: 28 April 2022).

Zeb, R. (2019). *Ethno-political Conflict in Pakistan: The Baloch Movement.* Routledge. https://doi.org/10.4324/9780429318139

Zurutuza, K. (2015) 'Understanding Pakistan's Baloch Insurgency', *The Diplomat,* 24 June. Available at https://thediplomat.com/2015/06/cracking-pakistans-baloch-insurgency/ (Accessed: 25 March 2022).

6 Manifestation of the COIN Strategy in Balochistan (post-2013) and Three Pillars of Counterinsurgency

Introduction

After successfully promulgating the Sub-Conventional Warfare (SCW) doctrine in 2013, the very next phase for the Pakistani Army was the effective implementation of the doctrine to curb insurgency in Balochistan and thus provide safety to the China-Pakistan Economic Corridor (CPEC). It was not an easy job. There were two essential problems attached to it. First was the inculcation of the new COIN approaches within the rank and file of the organisation. This was tantamount to entirely changing the organisational culture of the Pakistani Army from planning to fighting counterinsurgency campaign with the prime focus on interagency coordination. The second critical issue was adding legitimacy to the campaign in Balochistan and the formulation of a comprehensive COIN strategy involving political and economic prongs. This entailed the proactive involvement of the Federal Government of Pakistan and the Provincial Government of Balochistan. Yet another critical issue faced by the Pakistani Army was the design of a comprehensive information campaign for perception building in Balochistan as it was pivotal in establishing control and influence over the population. The Pakistani Army had no prior experience or expertise in it. This perception building and management issue took on a renewed importance against the backdrop of the CPEC projects in Balochistan. The general perception of the population, particularly the Baloch people, was anti-CPEC (Meer, 2015), which was based on the insurgents' narratives (Kazmi 2018, Interview, 19 December). This chapter assesses how, when and to what extent the Pakistani Army was able to overcome these fundamental challenges. Kilcullen's (2006b) Three Pillars of Counterinsurgency Model (3PCM) is applied to understand the security, political and economic components of the counterinsurgency campaign in Balochistan after the promulgation of the SCW doctrine in 2013.

The main reason for using 3PCM is that it provides a framework for the evaluation of a "whole-of-government" counterinsurgency strategy keeping in view the background and situation of Balochistan. This model has two parts: one pertains to the elaboration of the "conflict ecosystem" that forms the environment for the twenty-first-century counterinsurgency operations, and the other is a framework for the whole-of-government counterinsurgency in that environment. The description of the first part of the model that is the "conflict ecosystem" that forms the

DOI: 10.4324/9781003413905-6

environment is very similar to the prevailing environment in Balochistan. Likewise, the three pillars framework for whole-of-government counterinsurgency (the second part of the model for inter-agency collaboration) is supported by the Pakistani Army counterinsurgency doctrine to a large extent (as discussed in Chapter 4). These two claims have already been thoroughly elucidated in the earlier chapters of the book. This is why the 3PCM provides the best analytical and conceptual framework for evaluating the efficacy of the Pakistani Army's counterinsurgency strategy in Balochistan.

Moreover, the 3PCM emphasises the economic development in the insurgency ridden area. Interestingly, in the case of the current round of counterinsurgency in Balochistan, the political aim is to provide safety to the CPEC, a multi-billion-dollar economic initiative. This political aim, in turn, guarantees economic development/ sustenance in the province. Hence, the economic aspect of the COIN campaign in Balochistan relates to the overall objective of the CPEC itself. Therefore, 3PCM is well suited to be used as an analytical framework to evaluate the efficacy of the COIN campaign in Balochistan while concurrently considering the security, political, and economic aspects of the campaign.

First of all, this lengthy chapter explains the broad-ranging information strategy of the Pakistani Army and identifies how the security, political and economic pillars of the counterinsurgency campaign in Balochistan (post-2013) were blended into the information strategy. This information strategy is ultimately based on the political aim of the counterinsurgency campaign in Balochistan, that is, safeguarding the CPEC projects from all threats (Global Times, 2022; Siddiqui, 2019; Xinhua, 2018a; Khan Abdul Wajid, 2018). The chapter provides an assessment of the effectiveness of the Pakistani Army's conduct of counterinsurgency warfare in Balochistan after conceptualising the SCW doctrine (2013). Furthermore, the chapter discusses the flexibility (or otherwise) of the adopted COIN strategies, considers the nature of changes introduced to the strategy during the conduct of counterinsurgency operations in the province and the outcomes they led to. The chapter also explores the security, political and economic measures taken as a part of the overall COIN strategy, the role of military therein, the synergy between these measures, coordination between the provincial and federal governments while implementing the COIN strategy in Balochistan. Lastly, the chapter seeks to distinguish between the *Good* and *Bad* COIN practices of the Pakistani Army in the light of COIN scorecard (attached as Appendix 6). It helps to gauge the effectiveness of the Pakistani COIN during the fifth round of Balochistan insurgency. In a nutshell, the chapter seeks to understand the difference and equilibrium between counterinsurgency strategy and its practical manifestation on the ground in Balochistan after formally enacting the SCW doctrine (2013) by the Pakistani Army.

Information – Base of Three Pillars of Counterinsurgency Model (3PCM)

Information is the base within Kilcullen's 3PCM on which all other activities depend. It is crucial because in the words of Kilcullen: "perception is crucial in developing control and influence over population groups" (2006b, p. 5). The substantive steps in the security, economic and political domains are critical and can yield significant effects during counterinsurgency campaigns if they are well integrated into the overall information strategy. The information strategy involves the understanding of the effects of the counterinsurgency operations/efforts, especially on the local population, the global audience and the insurgents. In the COIN environment, every action of the counterinsurgents transmits a message. The purpose of the information campaign is to consolidate these messages into a coherent information strategy for the COIN campaign. It includes an array of activities such as Information Operations (IOs), media operations, intelligence collection, and dissemination to the various agencies involved in COIN campaign and various countermeasures to the insurgent ideology, motivation, and propaganda. Unless a firm base of information is developed, the three pillars of counterinsurgency cannot be effective.

On the other side, effective Information and Psychological Operations (Psy Ops) and effective media handling were great voids, owing to the absence of a formal policy on the subject, throughout the history of the Pakistani Army's counterinsurgency campaigns. The SCW doctrine (2013, p. 28) emphasises that essential steps for shaping the environment include a well-articulated Information Operation (IOs) campaign to legitimise the application of the military's response, winning public support and confidence through perception management and effective media handling to achieve the moral high ground. Perception management of the local populace in particular and the national and international community in general stands out to be an essential facet of IOs. In the context of Balochistan, the IOs are conducted as Psy Ops. This practice of the Pakistani Army in Balochistan will be elucidated in the ensuing section of the chapter. Hence, IOs turn out to be a fundamental aspect of the COIN campaign both in Kilcullen's 3PCM and SCW doctrine (2013).

Information Operations (IOs) Synonymous to Psychological Operations (Psy Ops) in the Pakistani Army

The Pakistani Army's counterinsurgency doctrine lays great emphasis on IOs. IOs focus on all segments of society involved directly or indirectly (a neutral majority, pro or anti-state elements as well as the militants) with well-conceived narratives and counter-narratives, using different means and mediums. As per the SCW doctrine (2013), IOs employ the following core capabilities; military security and deception, psychological operations (Psy Ops), electronic warfare and support to public diplomacy. The study will only cover the Psy Ops as most of the operations conducted in Balochistan under the rubric of IOs concerns the Psy Ops core capability, though some of the experts believe Psy Ops are a separate activity. Abid, an Information, Communication & Technology Officer (ICTO) in the Pakistani Army Corps of Signals, highlights this fact:

In the Pakistani Army, the terminology IOs is often used as a euphemism for Psy Ops. Thus, the IOs, which is a broader term encompassing integrated employment of a wide array of capabilities for influencing the audience, is interchangeably used to discuss the Psy Ops activities. This equally holds true once IOs is discussed for influencing the population or perception management in the context of Balochistan.

(2022, Interview, 24 January)

Therefore, this section emphasises Psy Ops in assessing IOs in Balochistan by primarily focusing on how well the Psy Ops initiatives were designed for the target audience in Balochistan. Moreover, it will highlight the effectiveness (including the strengths and weaknesses) of the intrinsic messages and themes of these initiatives, namely their acceptance among the Baloch target audience.

Psy Ops in Balochistan

Effective Information and Psychological Operations and media handling had been missing in the counterinsurgency history of the Pakistani Army. Owing to this reason, the SCW doctrine (2013) lays high stress on this aspect. The importance of this crucial facet of the counterinsurgency campaign in Balochistan was not fully realised until as late as 2013. Hamid, a former employee of the ISPR, recalls:

In 2014, the then DG ISPR and later Commander Southern Command, Quetta, General Asim Saleem Bajwa recognised the importance of information campaign for the counterinsurgency in Balochistan broadly for two reasons. Firstly, to counter the propaganda and ideology of the Baloch insurgents by presenting counter-narrative to the population. Secondly influencing the people of Balochistan by highlighting the importance of the CPEC as the backbone for the revival of the country's economy in general and Balochistan's progress in particular.

(2018, Interview, 09 December)

Moreover, the former Chief of the Army Staff (COAS) General Raheel Sharif in one of his visits to the Headquarters Southern Command, Quetta in 2015 is reported to have emphasised the need of convincing the Baloch audience that they were better off with the Pakistani Army in the province. The key message was that the future of Balochistan and the country lay in the successful implementation of the CPEC (Asim 2019, Interview, 25 January).

From 2014 onwards, the IOs (Psy Ops) have been taken up as a serious business by the Pakistani Army in Balochistan to promote the state narrative, in line with the SCW doctrine (2013). ISPR takes the leading role in collaboration with the Southern Command, Military Operations (MO) Directorate, Military Intelligence (MI) Directorate and the Federal Government. "The Provincial Government of Balochistan is also taken on board in this drive of information campaign but for financial and logistic purposes only", maintains Ajmal (2018, Interview, 17 December), a Public Relations Officer (PRO) at the provincial assembly of Balochistan.

Psy Ops are typically conducted to influence and inform the target audience during both conflict and peacetime (US Department of the Army, Psychological Operations, 2005, p. 1). The Psy Ops missions are divided into three categories, namely, Strategic, Operational, and Tactical Psy Ops. The Pakistani Army mainly resorts to the latter two types of Psy Ops in the context of Balochistan while keeping Strategic Psy Ops out of focus (Abid 2022, Interview, 24 January). Relegating the strategic Psy Ops is counterproductive as it facilitates the insurgent leaders in exile to influence foreign attitudes, perceptions, and behaviour through their information campaigns in the United Kingdom, Switzerland, and Germany. Moreover, the international audience remains ambivalent to the Pakistani state narrative on the counterinsurgency in Balochistan in the backdrop of CPEC (Ahsan 2019, Interview, 16 January). For example, during the ICC World Cup 2019 matches, a fly-past plane trailed political messages over the stadiums which read "Free Balochistan" and "Justice for Balochistan" (Headingley cricket ground on June 29 during a World Cup league match between India and Pakistan) (Lambert, 2019; Menezes, 2019) and "World must speak up for Balochistan" (during England versus Australia semi-final over Edgbaston on 11 July) (Business Standard, 2019). This was part of the IOs campaign organised by Baloch activists in Europe (Business Standard, 2019) to elicit awareness and sympathy from the international community.

Psy Ops – Means of Dissemination by the Pakistani Army

The primary means of Psy Ops dissemination in Balochistan are comprised of radio broadcasts, Internet (social media), press releases, leaflets/posters, and face-to-face communication with the local population in the form of Jirgas, etc. (Abid 2022, Interview, 24 January).

Radio

Radio is one of the essential means of disseminating information and thus influencing the minds of the population. The low literacy rate in Balochistan – 46 per cent (The Nation, 2018) is one of the big reasons for radio being an important means of dissemination in the province. At present, there are 15 different radio stations (both AM and FM) working in six districts in Balochistan (see Appendix 7). Almost all these districts are Baloch dominated areas except Quetta. Except for Radio Pakistan, which is the state entity, all others are commercial radio stations[1]. Ali, a local Baloch trader in Quetta and active political member of the Baloch National Party (BNP), highlights:

Most of the Baloch dominated districts are comparatively backward because of the security situation, so the majority of the people rely on transistor radio sets to remain abreast with the latest situation of the district/province. The Chinese made handheld transistor radio sets are very cheap and affordable for the poor people as they can buy it in less than 5 US$.

(2019, Interview, 13 January)

The Pakistani Army, through the Government of Balochistan, buys air time on these radio stations for public service announcements or provides press releases before launching any combat operations. Moreover, the talk shows covering and propagating the state agenda, in favour of CPEC, foster the image building of the military and expose the "lies" of insurgents[2]. These are designed and sponsored by the Provincial Government of Balochistan at the request of Southern Command/ISPR. Abid (2022, Interview, 24 January), an ICTO in the Pakistani Army Corps of Signals, confirms that almost all adult men listen to the radio broadcasts regularly. Badini (2019, Interview, 15 January), a former insurgent of BLA, also confirms that even the insurgent leadership listen to the pro-government content on the radio to plan their future course of action. It can be concluded that radio, in general, is an effective means of conveying messages in Balochistan.

Posters, Billboards, and Leaflets

The Pakistani Army display posters, panaflex banners, and large billboards aimed at undermining support to the insurgency at public places like crowded bazaars. The themes in favour of the CPEC, image building of the army and various other patriotic slogans are written on these. The Psy Ops personnel ensure that some graphics or pictures illustrating the theme or message should also be there along with the text on the posters – another by-product of the very low literacy rate. The same practice is followed while designing and distributing leaflets. Posters, leaflets, and billboards are taken as government/military propaganda by many people, depending upon the message they communicate. However, it can be said based on anecdotal evidence presented by the military personnel linked to Psy Ops in Balochistan that these help in reinforcing anti-insurgent sentiments in the segment of the Baloch society "sitting on a fence" wondering which side they should support: the insurgents or the army. Mostly the Government of Balochistan's financial and logistic resources are used for such initiatives.

Newspapers and Magazines

The Pakistani Army, in collaboration with the Directorate General Public Relations, Government of Balochistan also uses local and national newspapers and magazines for disseminating various themes and messages. This practice has been enhanced since 2014 as a result of ISPR's broader and active role for perception management as per the SCW Doctrine (2013) (Abid 2022, Interview, 24 January). It is important to note that all the famous national Urdu and English daily newspapers, like Dawn, Jang, Tribune, Nawa-i-Waqat, etc., reach even far off places like Turbat and Gwadar in Balochistan, albeit a day after publication. Moreover, Nawai Watan, the only Balochi language newspaper in publication, is also used for propagating the themes. Many of the stories and articles based on different Psy Ops themes like the role of CPEC in the economic uplift of Balochistan, project the positive image of the army and military recruitment advertisements are also published in these newspapers/magazines. Additionally, special editions are published on national days like 14 August, Independence Day and 6 September the Defence Day of Pakistan highlighting the positive image of the military busy in safeguarding the interests of the people of Balochistan by protecting the CPEC. The same content is also published on the digital editions of the established newspapers. The Provincial Government bears the financial expenditure of all such exercises (Ajmal, 2018, Interview, 17 December). The intelligence personnel sometimes, especially in the insurgent-dominated districts, also distribute these newspapers free of cost to shops, schools/colleges, restaurants, and to crowds as well.

It is essential to clarify how the use of print media is effective, keeping in view the high illiteracy rate in Balochistan, especially in rural areas. In Baloch society, those who are literate generally enjoy a superior status like a tribal chieftain, and they are capable of exerting disproportionate influence over the society. Many studies have shown that those engaged in terrorism or insurgency, especially the leadership, tend to be more educated[3]. This holds true for Baloch insurgents as well. For example, Dr Allah Nazar, the commander of the Balochistan Liberation Front (BLF), holds a Bachelor's degree in medicine. Abid (2022, Interview, 24 January) maintains that during Jirgas, he finds that most of the tribal elders are well conversant with the latest security/political updates of the province and fond of reading the newspapers. In Baloch culture, it is a common practice for those who can read to orally pass on the contents from the newspapers to the ones who cannot read. This is carried out, especially during the daily evening congregation of elders and young men in the villages as part of the social routine. The fact that print media is influencing the minds of the population in Balochistan is also a big concern for the Baloch insurgents, which is why they have tried to disrupt the circulation of newspapers/magazines time and again in the conflict-ridden districts (Kermani & Mudassir, 2017; Sameed, 2017). For example, in December 2017 and early 2018, the Baloch insurgents stopped the circulation of all the newspapers in more than a dozen districts of Balochistan by harassing the newspaper vendors and hawkers. Earlier in October the same year, "the BLF issued an angry ultimatum to local journalists [in Balochistan], whom they blamed for collaborating with the media wing of the Pakistan Army [ISPR]" (Kermani & Mudassir, 2017). It can be safely stated that print media in Balochistan is an effective means of disseminating the Psy Ops themes.

Internet and Social Networking

The internet services are generally poor in Balochistan and the 4G services for mobile users only exist in a few cities like Quetta, Gwadar, Sibbi, Khuzdar, and Zhob. Despite this, both the Pakistani Army and insurgents use social networking sites like Facebook, YouTube, Instagram, and Twitter to generate support for their respective cause. ISPR is covertly sponsoring a large number of Facebook pages/groups, video bloggers, and Twitter accounts to promote the Psy Ops themes of the military and counter the insurgents' ideology and narratives. Sometimes these proxy accounts engage in coordinated inauthentic behaviour[4] as part of a network, violating the terms and conditions of the social networking sites. In April 2019, 103 pages, groups, and accounts on both Facebook and Instagram, linked to ISPR, involved in this behaviour were removed by the site administrators (Gleicher, 2019). Facebook removed them because they attempted to conceal their identities by using fake accounts, to misrepresent themselves (Gleicher, 2019).

Abbasi (2019, Interview, 12 January) maintains that it is crucial to harness the power of social networking sites as a primary source of disseminating information. It also helps to influence a segment of the target audience in Balochistan who rely on the internet (mostly youngsters residing in major cities) for seeking information rather than the print media. Thus, it helps in influencing the educated Baloch youth. It is also useful to counter Baloch insurgents' false claims, made regularly on social media that they have killed Pakistani Military personnel.

Face-to-Face Communication

The Pakistani Army resorts to face-to-face communication in the conflict areas with the tribal elders in the form of routine meetings/Jirgas, dinner, or lunches arranged on various religious and cultural festivals for disseminating the information/themes. Face-to-face communication opportunities with civilians are organised starting from the company deployment areas by the company commanders and go up to the Battalion, Brigade, Division, and Corps level by the respective commanders. The messages and themes already disseminated by radio, posters, and social media, etc., are reiterated formally and informally in these get-togethers on a routine basis. Moreover, the concept of "every soldier is an ambassador of the Pakistani Military's goodwill for Baloch"[5] is also sometimes manifested by the infantry patrols in the conflict-affected areas in the form of informal greeting exchanges and honouring the Baloch tribal culture and values. Face-To-Face communication tends to be very useful in Balochistan, keeping in view the centuries-old Baloch culture of hospitality and honouring the guest at one's place (Abid 2022, Interview, 24 January).

Apart from these, there are hosts of activities carried out by the Pakistani Army on a regular basis in coordination and support, mainly financial and logistics, to the Provincial Government of Balochistan. For example, help with organising the annual Balochistan Sports Festival, Sibbi Mela (festival), Balochistan Youth Festival and International Exhibition in Gwadar, i.e., Gwadar EXPO started in 2018. These festivals are held on a regular basis to present the positive image of Balochistan to the international community, especially the Chinese government and foreign investors to encourage further foreign direct investment (FDI).

Target Audience and Themes

Psy Ops messages and themes pivoted around Baloch's yearning for prosperity and peace are likely to be effective. Ali (2019, Interview, 13 January), a local Baloch trader in Quetta and active political member of BNP, confirms that many Baloch people want a peaceful and prosperous life. They wish that economic and social progress should prevail in the province.

The important point to highlight here is that the society in Balochistan is not homogenous as it is divided by tribe and ethnicity. This phenomenon has a direct bearing on Psy Ops' target audience selection and the analysis of the themes. The key audience for the Psy Ops in Balochistan is the ethnic Baloch population. The ethnic Baloch population constitutes almost 55.7 per cent of the province's population, the Pashtun population represents 36 per cent of the population, while the remaining 8 per cent is composed of smaller communities such as Hazara, Sindhi, Punjabi, Uzbek, Turkmen, Tajik, and Saraiki (The Balochistan Post, 2023). The ethnic Baloch population mostly reside in those districts where the insurgents are active, and these areas are mainly along the route of the CPEC. It is essential to consider how well the Psy Ops were tailored for the target audience, that is primarily the Baloch ethnic people who are dominant in the province, especially in the conflict-affected areas.

As per the SCW doctrine (2013), the themes and message of the Psy Ops employed by the military should be consistent with the broader national security objective(s). In the case of Balochistan, specifically, after the MoU for CPEC was formalised in 2013, the state narrative of Pakistan is loud and clear that it needs to "safeguard the CPEC projects from all threats" (Global Times, 2022; Siddiqui, 2019; Xinhua, 2018a). It has also been set as the national security objective. This has been reiterated by both the civilian and the military leadership of Pakistan at various national and international forums.

The Psy Ops themes also indicate a potential change in the COIN strategy and the mindset of the Pakistani Army after the promulgation of the SCW doctrine (2013). In general, however, there is a significant continuity in military themes and messaging in Balochistan for the past five years or so. The underlying Psy Ops themes used by the Pakistani Army in Balochistan over the years are (Abid (ICTO), 2022; Interview, 24 January):

a CPEC brings peace, progress, and prosperity in Balochistan.
b Pakistani Army is the "Messiah Army" for the people of Balochistan.
c The Army requires you – so Baloch people should join the army and safeguard the CPEC.
d Baloch insurgents are the enemies of Balochistan and Baloch people.
e Amnesty and monitory rewards are offered for turning in weapons.
f The army will safeguard the CPEC project from the corrupt political elite and bureaucrats.
g A majority of the Baloch people are peace-loving, so they should support the Army.

Tracing the exact chronological development of specific Psy Ops campaigns in Balochistan is not possible because the central repository for any such data regarding themes and messages circulated or their effectiveness/impact on the target audience does not exist within the army. They do not keep formal records of the Psy Ops campaigns (Abid 2022, Interview, 24 January). It is also difficult to precisely and systematically test the target audience reaction to a specific Psy Ops theme because of the lack of opportunity for conducting surveys or focus groups by the researcher owing to security reasons in the insurgency affected areas of Balochistan. Keeping this in mind, the classification of the effects of the themes on the target audience as effective, partially effective, and ineffective are not based on a quantitative or mechanistic formula. The judgements of the effects of themes are based on the following: Interviews with Pakistani Military personnel who have interacted with the Baloch Jirgas, from time to time, and informally carried out local public-opinion survey polls of the tribal elders as a part of routine activity; interviews with people from various segments of the society in Quetta city (Balochistan); and open-source press reporting. Each of the themes mentioned above will be thoroughly discussed in turn, for assessing the effects on the target audience in Balochistan, in the subsequent section of the chapter.

CPEC Brings Peace, Progress, and Prosperity in Balochistan

This theme is an overarching one propagated by the Pakistani Army. The main purpose is to soften up the hearts of the people for Chinese investment and counter the insurgents' narrative that CPEC amounts to the economic colonisation of Balochistan. The military is using several means, including paid advertisements on national television to the use of print and social media to highlighting the benefits of CPEC for Balochistan. However, the theme is not making sufficient headway either amongst the population or the representatives of the population (legislators from Balochistan). There are several reasons for this. On the ground, facts do not suggest it to be true that CPEC is the recipe of progress and development for the people of Balochistan. For example, Balochistan gets a meagre share (less than 5 per cent) of CPEC investment.

This is what the former Chief Minister (CM) of Balochistan, Jam Kamal, complained about in the assembly: "Balochistan's current share under CPEC is only 4.5 per cent …… If we don't include Gwadar Port and HUBCO[6] projects, then Balochistan's share in CPEC is reduced to a meagre one per cent" (Pakistan Today, 2018a). The total investment portfolio of CPEC is around $62 billion. Out of this enormous investment, the improvised province of Balochistan gets a fraction, only $400 million, which is mainly for the development of the Gwadar Port neglecting other important development and energy projects. Moreover, Balochistan got only one energy project out of total 14 completed energy projects worth over $33 billion (CPEC Authority, 2022). The share of Balochistan is nil in the 13 electricity grid stations projects of 500 kv electricity transmission lines even though 88 per cent of the population is deprived of electricity (Zubair, 2019).

This situation is equally realised by all segments of society in Balochistan. The Balochistan assembly unanimously passed a resolution, in December 2018, demanding a due share in CPEC. The local population also realised the disproportionate distribution of the CPEC investment. Arif, an educated Baloch trader and political worker, maintains:

> We [people of Balochistan] know it well that CPEC will not be able to change our fate despite the strategic location of Gwadar Port, a lynchpin of the project. The primary beneficiaries of the project will be the people of Punjab and Sindh. The people of Balochistan will keep on swirling in the extreme poverty and militant attacks.
>
> (2018, Interview, 21 December)

This observation was confirmed by late Sardar Nasir, a tribal chief, a seasoned politician and a former parliamentarian for over four terms, in the following words:

> The local politicians of the province, many amongst them are also tribal elders/chiefs of their areas, know their people's concerns about the Chinese Investment projects. The biggest worry for the people of Balochistan is the disproportionate and ambiguous criterion of balance between benefits and burden of payments, vertical and horizontal distribution, for CPEC amongst the provinces. The unfair distribution of benefits and burden will only result in further deteriorating the condition of the already impoverished population of the province.
>
> (2019, Interview, 10 January)

The theme of CPEC as the beacon of prosperity for Balochistan has not gone down well. The situation, as mentioned above, strengthens the Baloch insurgents' claim that CPEC is the economic colonisation of the Balochistan and amounts to the plundering of the resources of the province. It can be fairly classified that the Psy Ops theme of CPEC brings peace, progress, and prosperity in Balochistan is ineffective at large.

Pakistani Army Is the "Messiah Army" for the People of Balochistan

The Pakistani Military's media activities have enhanced significantly since the upgrade of the ISPR's role and reorganisation in 2013 in light of the SCW doctrine. One of the principal themes, trailed by the ISPR, is the theme of the Pakistani Army as the "Messiah Army" (Shams, 2016). This theme has assumed new importance, particularly in Balochistan after the formulation of the MoU for CPEC, maintains Hamid (2018, Interview, 09 December). Unlike the previous theme, which was devoid of supporting facts and figures, this theme has substantial grounds. There is a long list of development projects carried out by the army. These range from establishing schools, colleges, hospitals, dispensaries, free medical camps, infrastructure development, water and sanitation systems and irrigation schemes. Some of these projects were undertaken solely by the army and others in collaboration with the provincial and federal governments. These development initiatives will be thoroughly discussed later in the chapter under the Political and Economic Pillar headings. These development initiatives, initiated primarily to win the hearts and minds of the local population, are well disseminated through electronic/print media, social media, banners, and billboards and most importantly during face-to-face communication at various levels by the army in Balochistan. Asif, a local journalist from Mastung district, believes:

> The Army's initiatives like establishment of school/colleges, setting up of water purification plants in the far-flung areas of Balochistan are taken very positively by the people. The local population has lost confidence in the Provincial Government as the government officials are corrupt and incompetent. Therefore, many people aspire that all the development works in the province should be done under the vigilance of the Army. Such initiatives for the welfare of the people of Balochistan definitely earn goodwill for the military.
>
> (2018, Interview, 07 December)

In addition to the development/welfare initiatives of the army, the ISPR had been regularly briefing on the operational outcomes of Operation Radd-ul-Fasaad[7] through various press conferences, press releases, tweets, and news flashes in the media. Shazia Haider, Quetta based female human rights attorney belonging to the ethnic Hazara community, highlights:

> The news of the killing of terrorists in Balochistan is always taken in a very positive spirit by the general population of the province. For example, we owe our thanks to the Army and the Frontier Corps (FC) for killing Salman Badini, one of the most wanted terrorists of Balochistan involved in the recent killings of over 100 members of the Hazara community.
>
> (2019, Interview, 11 January)

The rescue and relief efforts of the Pakistani army in 2022 during the devastating floods in Balochistan[8] were highly commended by the people of Balochistan, the international community (Business Recorder, 2022) and the national leadership of Pakistan alike (Radio Pakistan, 2022). Moreover, the people and the Provincial Assembly of Balochistan paid rich tribute and lauded the great sacrifice of Lieutenant General Sarfaraz Ali (Commander Southern Command) and five other military officials who lost their lives in a helicopter crash during flood relief operations in Balochistan on 2 August 2022 (Arab News, 2022; Daily Quetta Voice, 2022). By all such actions, the military has promoted itself as the saviour of the people of Balochistan, especially after a recent landmark change in its policy about the missing persons.

The Army Requires You – so Baloch People Should Join the Army and Safeguard the CPEC

This theme is spread for increasing the recruitments in the military and paramilitary forces keeping in view the enhanced security employment for CPEC. For providing the security cover to the CPEC and the workforce employed on this project, the Pakistani Army raised a Special Security Division (SSD) in 2016 (Raza, 2016). SSD is comprised of nine composite infantry battalions and six civil armed forces wings (with an estimated overall strength of the SSD to be 13,700 personnel) at the cost of around $220 million (Basit, 2019, p. 706).

The Pakistani Army conveyed this recruitment drive extensively to the people of Balochistan as elsewhere in the country. In Balochistan, the prime target of recruitment are Baloch ethnic people but not only limited to them. The army runs its advertisements very frequently through newspapers, posters, the internet and banners in the farthest areas of the province. Moreover, the mobile recruitment teams of the military and civil armed forces visited the far-flung areas to recruit the Baloch youth at the doorsteps. The mobile teams were a great help as the total area of the province is 347,190 sq. kms (Khan, 2009, p. 1074) and it was challenging for the people to reach the military recruitment centres at the district headquarters in the absence of good and frequent public transport. The figures shared by Fahad, Army Selection, and Recruitment Officer, Quetta (2022, Interview, 29 January) suggested that the recruitment rate, in comparison to 2015, had gradually increased by 12 per cent in 2016, 18 per cent in 2017 and finally reached up to 27 per cent increase in 2022 in the Baloch dominated districts[9]. This increase in the recruitment pattern is reconfirmed by Qasim (2022, Interview, 03 January), who is a journalist covering the districts of Turbat and Gwadar, though he had no specific statistics. He maintains that job security in the military attracts the Baloch youth. Moreover, the military has relaxed, as an exceptional measure, the age, qualification, and a few of the physical requirements for the Baloch people (Fahad 2022, Interview, 29 January).

The increase in the Baloch recruitment for SSD is also substantiated by the fact that the general trend of Baloch recruitment in the Pakistani Army, other than the SSD, has also increased many times over the last few years. For example, in December 2017, the then COAS General Qamar Bajwa highlighted that "nearly 20,000 sons of Balochistan were serving in Army" (Associated Press of Pakistan (APP), 2017) while addressing a seminar, *Human Resource Development for the youth of Balochistan – Opportunities and Challenges* at Quetta. The figure included over 600 army officers, while 232 cadets were undergoing training at the Pakistan Military Academy (PMA), Kakul (Zafar, 2017).

It is reasonable to believe that the theme "The Army requires you" is an effective one as the recruitment trend is encouraging given the Baloch people's history of antipathy towards the Pakistani Army, especially during the Musharraf regime.

Baloch Insurgents Are the Enemies of Balochistan and Baloch People

This theme is based on countering the Baloch insurgent narrative that they are fighting the war for safeguarding the economic and social interests of the Baloch nation against tyranny. Exclusive talk shows are aired on local radio stations in the Balochi and Baruhi languages challenging this narrative. Psy Ops messages focus on particular acts of destruction perpetrated by the Baloch insurgents like the killing of poor labourers, planting bombs in crowded places, blowing up train lines, and thus killing innocent people and disrupting the general election process as well by blowing up the polling stations (Khan Wajahat, 2015). All these activities invite military retaliation in which the civilian population in the particular area of operations also suffers. That is how insurgents are portrayed as the biggest hurdle in development and peace in the areas.

Interviews with some of the tribal leaders and the returning Pakistani Army personnel from the field suggest that many people are now fed up with the extraordinarily prolonged round of insurgency spanning over almost fourteen years. A majority of the Baloch people, particularly the youth, show signs of resentment against the Baloch secessionist insurgents for making their lives worse. A significant rise in the recruitment into the Pakistani Army of the Baloch youth over the last five to six years also hints towards the resentment of the youth against the insurgents.

This message was effective for some of the insurgents as well. For example, Obaidullah, a former BLA commander of a contingent of around 90 men, told NBC in an interview that his tipping point came once he confronted his comrades over targeting the innocent local population (Khan Wajahat, 2015). He mentions, "what started as an idealistic political fight for his people's rights has deteriorated into gangs extorting, kidnapping and even raping locals" (Khan Wajahat, 2015). It can be fairly categorised that the message Baloch insurgents are the enemies of Balochistan and Baloch people is effective.

Amnesty and Monitory Rewards Offered for Laying Down Weapons

The offer of amnesty and monetary rewards for laying down weapons in favour of the government is a message that was first disseminated in 2014 as a result of a reconciliation process kicked off by the military in consultation with the former chief minister of Balochistan Dr Abdul Malik Baloch (The Frontier Post, 2018). It continues to this day (refer Figure 6.3). This campaign has produced some very positive results, so it is worth pursuing (Durrani 2022, Interview, 11 August). This theme is assessed as effective.

The last two themes, "f" and "g" are basically the subthemes of "a" and "c" respectively. Based on the interviews, it can be safely assumed that the effectiveness of the last two themes in the list is the same as that of their parent themes.

Security Pillar of 3PCM

Military Operations: Transition from Conventional Practices to Modern Doctrine

To fully understand the military component (operational strategy after 2013) of the counterinsurgency strategy in Balochistan, it is imperative to briefly recapitulate the history of the operational strength and magnitude of the counterinsurgent forces traditionally employed in the province.

Traditionally the strategy of "Crush Them" (Paul et al., 2013a, p. 44), much in line with the inherited organisational culture of "butcher and bolt" was historically followed by the Pakistani Army during the operations in Balochistan, East Pakistan, and FATA. During the fourth round of Balochistan insurgency (1973–78), which resulted in a "COIN Win", the Pakistani Military "employed overwhelming force to crush the insurgency" (Paul et al., 2013b, p. 355). The various Pakistani Army studies, operational plans, and the Unit War Diaries show that the minimum force used in the operations were two infantry brigades plus a battalion of the elite commando Special Services Group (SSG) along with Frontier Corps (FC) in direct support. Most of the time the strength of the operating force was an Infantry Division plus-sized force and at times two infantry divisions, fully supported by the Air force, army aviation, and the integral artillery and mortar regiments[10]. Eighty thousand troops were actively involved at the peak of the operations during the fourth round of the insurgency in Balochistan (Paul et al., 2013b, p. 361). These operations always resulted in a massive loss of human life and severe collateral damage.

This practice continued until 2016 even after the publication of the SCW doctrine in November 2013. The main reason was that inculcation of the new doctrinal culture took considerable time and the army, especially the traditional battle-hardened units like the Frontier Force (FF), were good at stymying the change. They believed in "their way of fighting a war" more than any doctrinal guidance. Adeel (2019, Interview, 08 January), an infantry company commander, recalls that during operations in 2015 in the Malaar area of Kolwaah in Awaran district, it was challenging for him to stop his men from excessive firing of the automatic weapons and mortars on the suspected insurgent's hideouts which were at times in built-up areas. Their aim was to kill the "enemy" at all costs without accounting for anything else like collateral damage in case the insurgents were hiding in the civilian built-up areas. Moreover, those insurgents who even surrendered during the operations were killed on sight by the infantry units fighting against them. An amnesty scheme, not as a written policy, was limited to those who surrendered prior to the commencement of the operations. This cult of aggressiveness was in line with the heavy-handed enemy-centric approach, which was the organisational culture of the Pakistani Army. This was the case at the tactical and sub-tactical level till 2016.

Even after the promulgation of the SCW doctrine (2013), there was not a significant change within the army while planning the operations, where the MO Directorate designed the operational plans primarily in consultation with the concerned Corps HQ and the relevant services directorates. During the tenure of General Raheel Sharif, who succeeded General Kiyani as COAS, operations in Balochistan were planned while employing the maximum combat strength of the Southern Command that could be made available. The Operational Orders issued during that time by the MO and the Corps HQ show that usually no less than an infantry division-sized force with additional fighting and logistic attachments were deployed. This was no different than what was planned for FATA during General Raheel's tenure. For example, more than two infantry division-sized forces along with the substantial air support was used against Tehreek-e-Taliban Pakistan, and

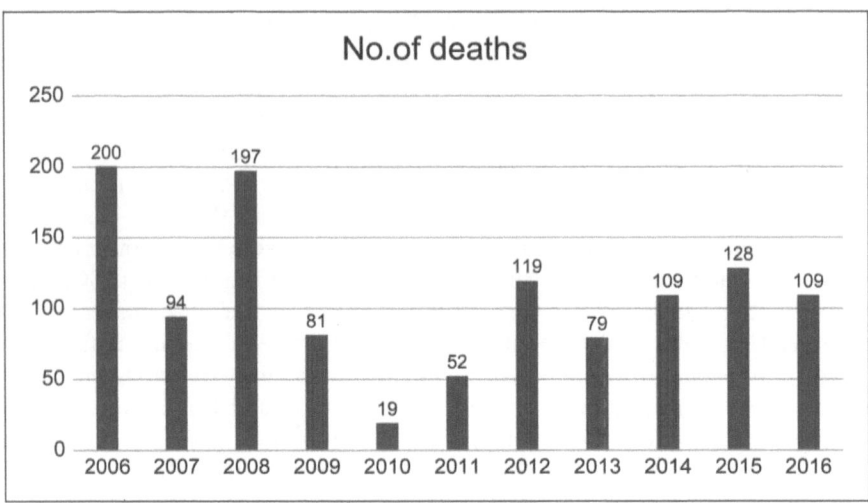

Figure 6.1 Number of Deaths as a Result of the State-Based Violence (2006–2016).

Source: Uppsala Conflict Data Program, Department of Peace and Conflict Research. Available at https://ucdp.uu.se/#conflict/325 (Accessed on 15th January 2023).

Al Qaeda linked militants in Operation Zarb-e-Azb in North Waziristan during 2014 (Adnan 2019, Interview, 14 January). In the same way, the operations in Balochistan were planned with massive force and heavy contingents of artillery, mortars, tanks, and air power. This led to a high number of deaths as was the case during the pre-SCW doctrinal years (2006-12) of the fifth round of insurgency in Balochistan (Nadeem 2018, Interview, 13 December).

Figure 6.1 shows the pattern of deaths resulting from state-based violence from 2006–16 during the fifth round of insurgency. The number of fatalities was highest in the initial years while dipping during 2010 and 2011 as the major focus of the operations of the army during these years was not Balochistan but FATA, against Tehreek-e-Taliban Pakistan (TTP) and foreign militants. The figure also highlights that after 2013, the number is not significantly less despite the promulgation of the doctrine. This shows that the lack of adoption to the new doctrinal themes on the ground at the tactical level by the traditional military units and following the same old planning parameters for operations at the army level contributed to the excessive use of force and led to a high number of killings.

There are a few explanations, apart from the organisational culture, for employing excessive force during operations as per the MO planning. The foremost factor was the lack of intelligence and overall limited Intelligence, Surveillance and Reconnaissance (ISR)[11] capabilities of the Pakistani Army due to a shortage of modern equipment for this purpose. It was a big disadvantage as the military was mainly relying on Human Intelligence (HUMINT). The intelligence gathering was, at times, not accurate, and actionable. Thus, the requirement for actionable intelligence, one of the core component of counterinsurgency warfare, which cuts across the entire spectrum of operations (Clark, 2006, p. 2) was missing. Moreover, the intelligence agencies were following the old practice of abducting people from Balochistan on suspicion of being insurgents, their collaborators, or sympathisers used to collect intelligence and information (Asim 2019, Interview, 25 January). Some of these persons are still missing after a decade (The Balochistan Post, 2019). This issue of the missing persons in Balochistan was a big counterproductive step taken in connection with the intelligence-gathering procedures. The issue of the missing person was taken very seriously by the general public in Balochistan because the abducted persons were kept in custody for long periods and tortured by the interrogating personnel belonging to the intelligence agencies. At times some of these persons died in the custody of the intelligence agencies owing to extreme torture. This practice aimed at extracting valuable intelligence from the alleged insurgents, their collaborators, or sympathisers brought an intense resentment against the army in the general masses, particularly the ethnic Baloch (Asim 2019, Interview, 25 January). Another factor which further aggravated the situation was the lack of coordination in the acquisition and dissemination of the intelligence amongst the various intelligence agencies operating in Balochistan. There was no formal mechanism for this purpose in existence.

This lack of actionable intelligence gathering and timely coordination and dissemination led to the massive force employment during the operations in Balochistan, which was ultimately a force dissipation. Had there been accurate and actionable intelligence, this massive employment of force could have been avoided (Ahsan 2019, Interview, 16 January). As in the words of Hoffman: "Essential to the effective application of that force is the acquisition of actionable intelligence, its rapid and proper analysis, and, perhaps most critically, its efficacious coordination and dissemination" (2004, p. 10).

The net result of the massive force employment was huge collateral damage. This ran counter to the newly enacted SCW doctrine (2013) of the army. The SCW doctrine reads:

> Indiscriminate use of firepower, which causes collateral damage to be avoided. Use of air and artillery must be selective and effective .... Collateral damage can be reduced through accurate, actionable/timely intelligence and selective but effective use of force. As a principle, no collective punishment is to be meted out.
>
> (2013, pp. 36–37)

In December 2016, the 197th Corps Commander Conference was chaired by the then new COAS General Qamar Bajwa. One of the top agenda points was the change in operational strategy of the military, particularly in Balochistan, in light of the SCW doctrine (2013) and how it could be implemented in true letter and spirit. Ahsan (2019, Interview, 16 January) maintains that there were a couple of reasons for it. Firstly, General Bajwa was a member of the board, constituted by General Kiyani during his tenure as COAS, for developing the SCW doctrine (2013). Therefore, he was keen to implement it. Secondly, and the most pressing reason was the Chinese concerns over the security situation in Balochistan with regards to the implementation of the CPEC. After the 197th Conference followed by several in-house discussions (IHDs) at the MO directorate, it was decided that the strategy of Intelligence Based Operations (IBOs) coupled with or immediately followed by an appropriately scaled combat component of the military will be adopted at large in Balochistan.

Intelligence Based Operations (IBOs)

There were two cardinal operational aspects of the IBOs. The first was the need for sophistication in accurate, timely and actionable intelligence collection by relying on and enhancing expertise/equipment for HUMINT, Geospatial intelligence (GEOINT), and Signals intelligence (SIGINT). Moreover, it was equally important to develop the operational procedures for coordination and timely dissemination of the intelligence information between military and civil intelligence agencies in Balochistan. This aspect goes well with the SCW doctrine as it highlights:

> The operations in sub-conventional warfare are mainly intelligence driven and continue through all stages of the conflict. It is imperative to have an integrated approach by all civil and military intelligence agencies while employing all methods of human and technical intelligence.
>
> (2013, p. 28)

Secondly, interagency coordination for appropriate, effect-based and focused use of force was set to be the hallmark of the IBOs. This was in line with the SCW doctrine which emphasises that: "Force used against the militants must be appropriately scaled, effect-based and focussed in the application. The disproportionate and wanton use of force may undermine public support at any stage of the operation" (2013, p. 28).

To enhance the military intelligence capacity, the surveillance equipment for the aerial platforms and sophisticated signals equipment were purchased by the Pakistani Army from China and Russia in 2017 as the sole available and affordable option (Muzammil, 2018, Interview, 27 December). Both officers and troops were sent on specialised and technical intelligence training to these countries, instead of the United States, keeping in mind the deteriorating political, diplomatic, and military ties with the United States which ended up in the Trump's administration cuts in 2018 on military training programmes for Pakistani officers (Ali & Stewart, 2018).

Though originally conceived for Balochistan, later it was decided that the IBOs will be launched countrywide as part of Operation Radd-ul-Fasaad to eliminate any threat of insurgency, militancy, and terrorism. Firstly, this was to give confidence to the Chinese authorities who were concerned about the security situation in Balochistan but were also sceptical about the situation in FATA/KPK because of the presence of TTP and the Uyghur militant group the East Turkestan Islamic Movement (ETIM). China had also earlier urged General Raheel Sharif to weed out militants from Xinjiang, who were holed up in a lawless tribal belt in Pakistan (FATA), home to a lethal mix of militant groups, including the Taliban and al-Qaeda (South China Morning Post, 2015). It is believed that owing to Chinese pressure, General Raheel launched Operation Zarb-e-Azb in North Waziristan in 2014, which was a massive deployment of force as Jaffrelot maintains: "It [Operation Zarb-e-Azb] was possibly launched under pressure from China, which is increasingly fearful of Uighur militants, who are trained in the Federally Administered Tribal Areas (FATA)" (2014, para.7). Operation Radd-ul-Fasaad was launched all across Pakistan as an assurance to the Chinese government about the safety of the CPEC along with eliminating any residual/latent threat of ETIM.

Secondly, Radd-ul-Fasaad could be integrated into the already existing but partially functional National Action Plan (NAP)[12] jointly conceived in 2014 by the political and military leadership of the country. To this effect, the ISPR (2017a) press release No PR-87/2017 maintains that the "Pursuance of National Action Plan will be the hallmark of this operation [Radd-ul-Fasaad]". Operation Radd-ul-Fasaad was officially announced on 22nd February 2017 (ISPR, 2017a). The first-ever targeted operation as a part of Radd-ul-Fasaad was conducted by the Army, FC, and intelligence agencies in Balochistan as per the press release No PR-88/2017 dated 23rd February 2017 (ISPR, 2017b). The operation was successful as it averted a major terrorist incident by capturing 23 Improvised Explosive Devices (IEDs) along with the perpetrators (ISPR, 2017b). This shows that Balochistan had been the prime focus of the renewed strategy in the form of Operation Radd-ul-Fasaad. Moreover, the press release highlights that it was an IBO jointly conducted in line with the spirit of interagency coordination.

Operation Radd-ul-Fasaad's Manifestation on the Ground, SCW Doctrine, and Effects in Balochistan for Securing the CPEC

Legitimacy

The most significant advantage the Pakistani Military attained by proclaiming the Operation Radd-ul-Fasaad in pursuance of the NAP was that the operation gained legitimacy. According to many, declaring it as a part of NAP was very timely and intelligent move (Sardar Nasir 2019, Interview, 10 January; Ahsan 2019, Interview, 16 January). The utility of this step, in terms of legitimacy, is captured in the following statement of Sardar Nasir, the tribal chief, a seasoned politician and former parliamentarian from Balochistan:

It was a wise move of General Bajwa to declare Radd-ul-Fasaad as a continuation of the National Action Plan (NAP) – 2014 though the latter was practically almost non-functional at the time of the former's launch. This saved the military's time and efforts to get it passed from the parliament as the parliament had already approved the NAP. Unfortunately, any legislation about Balochistan takes a lot of time to develop consensus and most of the time it is not passed because of the dirty politics and lobbying amongst the legislators and split between the upper and lower houses of the parliament. As Radd-ul-Fasaad was the part of NAP so nobody, at least not in my knowledge, raised any objection over it even in Balochistan.

(2019, Interview, 10 January)

Asif (2018, Interview, 07 December) agrees that it was a beneficial step to integrate Radd-ul-Fasaad with NAP. Though the Pakistani Military has a long history of conducting counterinsurgency operations without taking parliament into its confidence, the new realisation of the increased legitimacy of the operation by the military has yielded positive results in increasing public confidence in the military actions in Balochistan.

The legitimacy of the counterinsurgency is one of the vital facets of the 3PCM (Kilcullen, 2006b, p. 5). Moreover, the SCW doctrine (2013, p. 23) amply highlights the importance of legitimacy as it gains public support for the counterinsurgency strategy. Legitimacy is also derived from the perception that authority is genuine and uses proper agencies for rational purposes (SCW doctrine, 2013, p. 24). In the case of Operation Radd-ul-Fasaad in Balochistan, this perception was created through the already established legitimacy of the NAP and the use of an appropriately scaled combat component of the military in intelligence driven operations. Moreover, the development projects carried out by the Pakistani Army, such as establishing the cadet colleges and schools, water purification schemes, increased intake of Baloch youth in the army, DDR programs, etc., as part of the winning hearts and minds initiative in Balochistan has also helped in increasing the moral legitimacy of their operations. All these initiatives are discussed in detail later in the chapter. Alongside the de-legitimisation of the insurgents through another effective Psy Ops theme by the army "*Baloch insurgents are the enemies of Balochistan and Baloch people*" also contributed to the legitimacy of Operation Radd-ul-Fasaad. In short, all these factors collectively helped the Pakistani Army in winning political and moral legitimacy in Balochistan.

Insurgent Approaches and Responding COIN Practices

The various Pakistani Army studies, operational plans, and the Unit War Diaries show that the approach adopted by the insurgents has not changed significantly; except the use of suicide attacks, better intelligence gathering and deception. The insurgent approach is based on guerrilla tactics. They usually resort to raids and ambushes in a platoon size strength, that is between 36 to 45 men depending upon the terrain condition and the target, further divided into the vanguard, action group, and the rear-guard groups. The targets of ambushes are normally the military and LEAs convoys. At present, the convoys carrying Chinese nationals run a very high risk of being ambushed by the insurgents irrespective of the heavy security surrounding them. Raids are also carried out frequently depending upon the importance of the target. Insurgents frequently raid CPEC related installations and buildings and accommodation where Chinese and Pakistani officials working on CPEC are based. For example, the insurgents stormed a luxury hotel in Gwadar in May 2019, where they suspected the Chinese engineers were staying (Safi & Baloch, 2019). Baloch insurgents also attacked the Chinese consulate in Karachi in November 2018. More recently, in April 2022, a female BLA militant carried out a suicide bombing near Karachi University's Confucius Institute, killing three Chinese nationals (language teachers) and their Pakistani van driver (BBC, 2022).

Apart from raids and ambushes, the insurgents also resort to heavy snipping at long range when the target was beyond their capability to ambush or raid, or as bait to invite the troops to move into the killing area where an ambush was already laid. The insurgents set up a broad network of base camps usually in the barren mountains traditionally and commonly known as *Parari*[13] *Camps*. These Parari Camps vary in size and could hold from 36 to 200 armed men along with automatic weapons, including rocket launchers, machine guns and mines, etc. These armed men carried out insurgent activities in their designated area of operation under the guidance and order of the insurgent commander of the camp who usually gets covert patronage from some local Sardar (tribal chief) of the area. Currently, there is a slight change in the insurgent strategy to avoid complete annihilation by the attack helicopters and aerial bombardment of the jets. Now almost ninety per cent of the insurgent force of each base camp distribute into various armed mobile teams and establish their hideouts in the civilian built-up areas to minimise the chances of detection and heavy-handed operation by the military (Badini 2019, Interview, 15 January, a former BLA insurgent). The base camps are manned by a very small number of men. This strategy is consistently followed. These armed mobile teams are still linked to their base camps for the provision of weapons and ammunition and planning, etc.

In response to the renewed insurgent guerrilla strategy in Balochistan, the Pakistani Military mostly resorts to intelligence driven Broad Spectrum Security Operations, named Radd-ul-Fasaad as mentioned earlier. There are specific characteristics of these operations which are in line with the SCW doctrine (2013).

Interagency Coordination

Operation Radd-ul-Fasaad entails the coordinated employment of the Pakistan Air Force, Pakistan Navy, LEAs, and intelligence agencies active participation to support the army's efforts to eradicate insurgent and militant threats (Shah Kriti, 2017). The IBOs are conducted after thorough intelligence gathering while employing methods of human and technical intelligence. The HQ Southern Command (SC), Quetta is the coordinating HQ of these operations; in other words, it acts as a nerve centre for operations in Balochistan. At HQ SC, the information is collated, organised, processed, and disseminated efficiently. The IBOs target the insurgent hideouts which are mainly found in the populated areas. A composite force, from a platoon to an infantry battalion-sized strength depending upon the nature of sanctuary and expected resistance, made up of the military in a lead role and supported by the civil LEAs (mainly FC and police) is used to target the insurgents. Aerial surveillance is done through drones and Bell AH-1 helicopter. ISR capability of the

Pakistani army has enhanced multifold with the addition of a squadron (5 UAVs and two ground control units) of CH-4[14] multirole medium-altitude long-endurance (MALE) unmanned aerial vehicles (UAVs) acquired from China in 2021 (Janes, 2021; The EurAsian Times, 2022). The CH-4 is currently in service in Balochistan against the insurgents (The EurAsian Times, 2022). In this way, the unity of effort and unity of command principles are thoroughly practised in Balochistan (Qadeer 2022, Interview, 08 December).

The exceptions to the strength mentioned above (platoon to a battalion-sized force) are the operations conducted against the Parari Camps in mountains with a large confirmed insurgent presence. In these cases, an infantry division-size force is used in conjunction with close combat air support. These large scale "major operations" are very rare after the change in the insurgent tactics of dispersing their force in hideouts, permanently, within the population. For example, out of 1410 IBOs conducted in Balochistan during 2017, only 29 were major operations which makes just 2 per cent of all the operations as briefed by Major General Sahir Shamshad, the then DG MO, to the Senate Committee on National Security (Guramani, 2017).

Collateral Damage, IDPs, ROE, and Political Primacy/Ownership

The SCW doctrine's guidance is inherent in Operation Radd-ul-Fasaad. Unlike in the past, the operation has no coercive undercurrents. The force used is appropriate, effect-based and focused as per the SCW doctrine (2013), maintains Qadeer (2022, Interview, 08 December). The change in operational strategy from massive force to the appropriate, effect-based, and focused use of force in the form of IBOs is also reflected in Asad's statement (2019, Interview, 02 January) a battalion commander in Balochistan. He maintained that his battalion was busy in fencing the Af-Pak border in Balochistan because for nearly the last two years the force employment in Balochistan was not as high as it used to be earlier. Though the frequency of operations had not decreased, the force used during IBOs was appropriately scaled to avoid collateral damage. Therefore, the troops of the SC could be made available for the border fencing purpose. Moreover, the avoidance of excessive force, unlike in the recent past until early 2016, has reduced the collateral damage. For example, one form of significant collateral harm is the displacement of the population from the area of operation (Davidovic, 2018). In the earlier rounds of insurgency and even before Operation Radd-ul-Fasaad internally displaced persons (IDPs) were a common phenomenon to avoid unnecessary killings because of the excessive use of force as explained earlier in the chapter. Apart from IDPs, the loss of property, livestock, crops and businesses was very high owing to the operations in Balochistan. After the launching of Operation Radd-ul-Fasaad, all sorts of collateral damage have been reduced considerably, including the displacement of the population from the areas of operation. This also entails the pivot around which the Rules of Engagement (ROE) revolve in SCW doctrine, and that is all possible efforts are made to avoid collateral damage. Adhering to this mindset has led to a significant decrease in fatalities as a result of state-based violence by the counterinsurgents

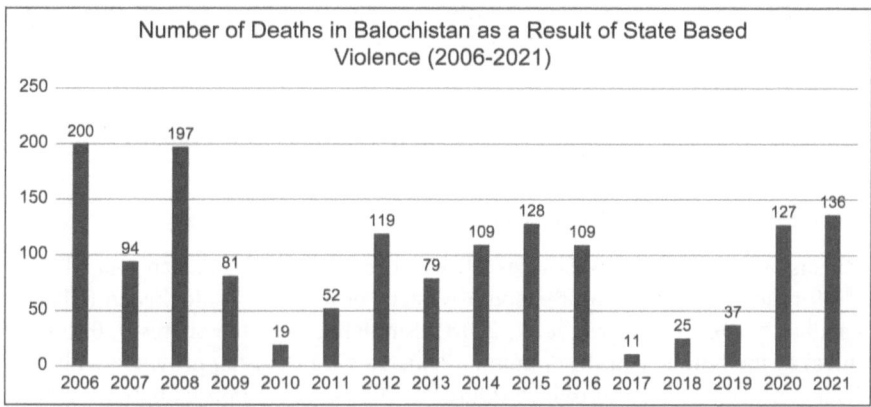

Figure 6.2 Number of Deaths as a Result of the State-Based Violence (2006–2021).[15]

Source: Uppsala Conflict Data Program, Department of Peace and Conflict Research. Available at https://ucdp.uu.se/#conflict/325 (Accessed on 15th January 2023).

after 2016 till 2019 only. Figure 6.2 (above) demonstrates the significant dip in the fatalities during 2017, 2018 and 2019 but again a sharp rise in death toll from 2020 to 2021 as a result of state-based violence by the Pakistani military/LEAs.

The prime reason for this increase in fatalities is linked to the presence and alliance formation spree of ISKP with the local insurgent organisation and sectarian militants in the province of Balochistan. From 2019 onwards, ISKP has established a strong foothold in Pakistan after alliance formation with BLA and LeJ, as mentioned in the prvious chapter. This alliance strategy has helped ISKP carry out attacks across the length and breadth of Balochistan, as mentioned in Chapter 5. This has led to an increase in attacks in Balochistan, especially against the CPEC targets, the Chinese nationals and the military/LEA personnel and installations in the province. The Pakistani military has no strategy to counter this alliance formation which is greatly facilitated by the porous border with Afghanistan despite the border fencing project, which is currently on halt and in shambles owing to numerous diplomatic, political, and financial reasons and cross border fire clashes as mentioned earlier in the book. This has severe implications for the smooth progress of CPEC in the province. The Pakistani military is devoid of any strategy to check the alliance formed between the local insurgent organisations, such as BLA, and the sectarian militant groups like LeJ and the international Jihadists like ISKP in the region. These alliances are about to transform into enduring, more potent and lethal networks with the ability to move resources, human resources and logistics across the borders of Pakistan, Afghanistan, and Iran.

The SCW (2013) is also silent on the subject of disrupting the nexus of local insurgents and militants with international jihadists. Therefore, the statistics show that as the frequency of violent attacks by the BLA and LeJ in collaboration with ISKP increased in the province during 2020 and 2021, so did the use of force by the Pakistani military. The military resorted to this pre-doctrinal practice sheer out of frustration amongst the rank and file regarding the relatively new phenomenon of international jihadists (ISKP) in the province and simply no line of action to disrupt these alliances. The security establishment in Pakistan needs to understand the causes of these alliances to capitalize on the various dimensions of disruption they present. The SCW doctrine (2013) needs updating on the subject.

Insecure Porous Borders with Iran and Afghanistan and Border Fencing

Pakistan shares nearly 1500-miles of the border with Afghanistan, out of which a 789-mile stretch is along the south-western province of Balochistan (Chaman-Kandhar border) (Yusufzai et al., 2018). Similarly, Pakistan shares 950 km (590 miles) of the border with Iran, linking Pakistan's Balochistan province with Iran's Sistan and Baluchestan province (Qureshi, 2019). These porous borders with the rugged mountainous and desert terrain facilitate the illegal cross-border movement of insurgents from Balochistan to Afghanistan (Lieven, 2017, p. 179) and in some cases to Iran as well. In the earlier round of the Balochistan insurgency, the insurgents fled to Afghanistan to avoid capture by the military and established their bases for operating in Balochistan as guerrilla fighters (Naseer, 2010; Paul et al., 2013b, p. 355). The same practice is still on-going. Currently, the external threat to CPEC originates from these porous borders which are used by militants and insurgents fleeing from the operations of the LEAs as well as for logistics and arms/ammunition supplies, training camps, bases, and as a launch pad for carrying out terrorist attacks in the province. In this connection, Islamabad has also blamed India for igniting the insurgent activities in Balochistan through the border of Afghanistan for fulfilling its ulterior goal of sabotaging the CPEC (Iqbal, 2018, p. 207). Moreover, Pakistan alleges that the militants of the Islamic State Khorasan (ISKP) also use the porous border between Balochistan and Afghanistan for planning attacks in Pakistan and accuses Kabul of not doing enough to secure the border (Yusufzai et al., 2018).

While realising the stark consequences of the porous borders, the Pakistani Military initiated the ambitious border security project, that is the chain-link fencing[16] of the complete Pakistan-Afghanistan (Af-Pak) border at the cost of $483 million (Yusufzai et al., 2018), in 2017, as part of Operation Radd-ul-Fasaad (ISPR, 2017c). In the first phase, the Pakistani Army had completed the fencing of the Af-Pak border in KPK and the erstwhile FATA region by December 2018. In the second round, fencing commenced on the Af-Pak border in Balochistan. In Balochistan, the HQ SC utilised all available manpower along with FC, around 1500 to 2000 troops, for the fence construction. This was the manpower which could only be made available as the result of the change in the operational strategy in the form of IBOs where a large number of troops were not committed. The ISPR claimed that the fencing would be completed by the end of 2019 (Gul, 2018), but it could not be completed. The fencing on the Af-Pak border in Balochistan had been extremely slow and is currently almost on halt. There are several reasons for it. Firstly, the severe existing economic crisis in Pakistan led to a decrease in the defence budget of the army (The Economic Times, 2022), adversely affecting the fence construction. Secondly, the frequent incidents of provocative firing earlier from the Afghan security forces and now Taliban forces on the military workers erecting the border fence has prevented the complete fencing of the Af-Pak border in Balochistan until now 2022.

As a second step, the Pakistani Military envisages the fencing of the 950 km Pakistan-Iran border (Qureshi, 2019). The construction of a fence along the Pakistan-Iran border has not yet commenced. The slow progress over the fencing in Balochistan prevents the military from disrupting the tangible support to the insurgents and ISKP militants from across the border. This is one of the major impediments in curbing the menace of insurgency in Balochistan which has a direct bearing on jeopardising the security of the CPEC in Balochistan.

Concerted Force for Physical Security of CPEC Sites, Infrastructure, and Personnel (Workforce)

In a challenging security situation for the CPEC, the Pakistani Army set up a Special Security Division (SSD) in 2016. It comprised of 9,000 Pakistani Army soldiers (nine infantry battalions) and 6,000 para-military forces personnel (six wings of FC) for the security of the project (Raza, 2016). This set up cost $220 million (Basit, 2019, p. 706). This is the total force for guarding over 2200 km network of roads, railways, and pipelines, special economic zones, power plants, Gwadar Port and all the affiliated infrastructure along with over 2000 Chinese nationals working on the CPEC from Gwadar to Khunjerab Pass in the north. Out of this total force, Balochistan gets a major chunk that is one infantry brigade (three battalions) and all the six wings of FC. Apart from this, 3000 policemen, 1000 Levies personnel, Pakistan Naval Marines Force, and border security forces will together guard all routes to the port of Gwadar (Basit, 2019, p. 706). The Pakistan navy has also enhanced its capacity to provide maritime security to the CPEC by establishing the Pak Naval Task Force 88 (NTF 88) which is fully active. The purpose of this elite maritime force is to protect Gwadar Port and linked channels. Apart from two naval vessels donated by China for the maritime security of CPEC, NTF 88 has frigates, naval aviation aircraft, gunboats, drones, and several ISR systems in its inventory (Basit, 2019, p. 706).

The existing security personnel strength of all military and LEAs in Balochistan is not adequate for the security of the CPEC/Chinese personnel, carrying out IBOs, providing human security and public safety. There is a great deficiency of the security personnel in the province keeping in mind the 134,050 sq. mi. area (nearly the size of Germany) of the province, constituting 44% of Pakistan's total landmass, the diversity, and scattered CPEC projects, the length of the CPEC routes, the on-going insurgency, sectarian militancy, and the presence of ISKP. There were reports in 2019 that the Pakistani Military was going to raise another security division in the near future (The Tribune, 2019) but it could not be raised till now. The primary reason for not raising another security division is the lack of finances which is a big question mark at the moment as Pakistan is still struggling hard to receive an IMF bailout package (Mangi, 2023) despite massive FDI for the CPEC.

In this security matrix of the CPEC in Balochistan, the most prominent missing link is the strength, training, efficacy, and equipment of the Balochistan Police and Balochistan Levies Force. In the security pillar, the main tasks of human security and public safety, especially in the urban centres lie in the domain of the police and in the rural areas in the domain of the Levies Force in Balochistan. Neither the Balochistan police nor the Balochistan Levies Force are professionally trained in counterinsurgency policing operations. Furthermore, the SCW doctrine (2013) is primarily focused on the role of military forces in quelling insurgencies and pay scant regard to the essential role that police and Levies play in pacifying an insurgency and maintaining long-term peace[17]. Moreover, both forces suffer from an acute shortage of manpower (Durrani 2022, Interview, 11 August). It is essential to make up the shortage of manpower in these two LEAs as adequate police [and Levies] strength, in both personnel and material, is a necessary if not a sufficient condition for effective COIN (Naseemullah, 2014). In May 2019, the Balochistan government approved a special security division of police with over 1305 vacancies for all ranks, but for Quetta District only (Dawn, 2019). However, there is no report till now about the source and allocation of funding for this mega initiative, so it is expected that this plan of the police security division will not materialise at all (Durrani 2022, Interview, 11 August). Likewise, the Balochistan government approved a comprehensive four-year "revamping and restructuring" program for Balochistan Leives Force in September 2018 to make it an efficient and modern force and to bring it to par with other LEAs. New specialized wings for the Security of CPEC, counter-terrorism, digital communication, intelligence and investigation, quick response force (QRF), and bomb disposal squad(s) were planned to be established (Balochistan Levies Force, 2021). Owing to the non-allocations of requisite funds for this "revamping and restructuring" program, the Balochistan Leives Force could not be scaled up to the desired level and still suffers from the inadequacy in manpower, equipment, and training for COIN operations (Durrani 2022, Interview, 11 August).

Political Pillar

Historically, the political process is essentially attached to the success or failure of the counterinsurgent force (Galula, 1964, p. 63). The first principle for countering the insurgency is that the government must have a defined and clear political aim to initiate a political process for establishing a united, politically and economically stable country (Thompson, 1966, pp. 50–52). In the case of Balochistan, the political aim is very clear in the state narrative, after 2013, and that is "safeguarding the CPEC projects from all threats" (Global Times, 2022; Siddiqui, 2019; Xinhua, 2018a). The objectives to achieve this aim comprise managing violence through political initiatives by marginalising insurgents and other militant groups and simultaneously offering reconciliation opportunities, extending governance, and promoting the rule of law by establishing the writ of the government in the province. The political initiatives and collaboration of the Federal Government of Pakistan and the Provincial Government of Balochistan are based on these aims and objectives.

Political synergy exists between the Provincial Government of Balochistan and the Federal Government on the security issue of Balochistan against the backdrop of the CPEC. For example, the former PM Imran Khan on his visit to Quetta in April 2019 expressed, in a meeting with the Balochistan CM and the provincial cabinet ministers, that his government is determined to implement NAP in the province in true letter and spirit (Radio Pakistan, 2019). Earlier in March 2019, during the apex committee of Balochistan meeting, the CM Jam Kamal vowed to implement the NAP in its letter and spirit across the province (Khyber News, 2019). This political synergy is in continuation even after the political regime change in April 2022 in Pakistan. For instance, in June 2022, during a meeting, PM Shehbaz Sharif and the CM of Balochistan, Mir Quddus Bizenjo, agreed to continue implementing the National Action Plan (NAP) to improve the law-and-order situation in the province and provide security for the ongoing development projects of the CPEC. Earlier the same day, the PM announced a 100 billion PKR development package for Balochistan during his day-long visit to Gwadar on June 24, 2022 (Shah, 2022b). More recently, in February 2023, the Provincial Apex Committee, during its 13th meeting held under the chairmanship of CM Balochistan Mir Quddus Bizenjo, emphasized again the strict implementation of the National Action Plan to provide foolproof security and safety to the CPEC in the province (Shahwani, 2023).

Like other pillars, the political pillar also develops with the principal dimensions of effectiveness and legitimacy. This political synergy between the Federal Government and the Provincial Government provides legitimacy to the counterinsurgency campaign in Balochistan.

The SCW doctrine (2013) promotes the idea that military or political solutions in isolation cannot defeat an insurgency as it requires a comprehensive approach with a blend of political and economic measures coupled with the use of legitimate armed forces. This is very much in line with Kilcullen's 3PCM.

Missing Persons Issue

The relative deprivation of Balochistan in comparison to other provinces inculcated a sense of insecurity and rebellion in the Baloch people and ultimately resulted in the various rounds of the insurgency in Balochistan. The issue of missing persons, mainly the members of the Baloch ethnic community, in the custody of the intelligence agencies is a major concern for the people of Balochistan. In 2018–19, the Balochistan National Party (BNP), a political coalition partner of the government, compelled the Imran Khan's administration to address the issue of missing persons in Balochistan. The federal government's efforts resulted in the intelligence agencies and military initiating the process of releasing missing persons. In mid-October 2018, approximately 300 missing persons were released (Alvi, 2018), and since then, a consistent stream of missing persons has been returning to their homes in Balochistan. For instance, in June 2019 alone, over 200 Baloch missing persons reached their homes (Shah Syed, 2019). According to the Central Police Office in Quetta, over 1,000 Baloch missing persons had returned by July 2019. This political action bolstered the counterinsurgent forces' goodwill in Balochistan by releasing missing persons in large numbers, countering the insurgent narrative of state-perpetrated atrocities. Additionally, it has contributed to enhancing the military's image as the saviour of the people of Balochistan.

Unfortunately, the situation in Balochistan changed after Imran Khan's government was ousted in 2022 through a vote of no confidence in the national assembly. The intelligence agencies have reverted to their old tactics of forced disappearances of people suspected of being insurgents or collaborators/facilitators of insurgents, followed by extrajudicial killings of some in captivity. For example, since the fall of Imran Khan's government in 2022, 584 people have gone missing, and 59 individuals have been extrajudicially killed in Balochistan, according to the Baloch Human Rights Council (2022, 2023). Although the international press reports even higher figures, with 629 individuals forcibly disappeared and 195 killed (ANI, 2023). Though these statistics appear to be over-exaggerated, but it is clear that the intelligence agencies and law enforcement authorities have once again resorted to their decades-old practice of forced disappearances and extrajudicial killings in the province.

However, the current government of the Pakistan Democratic Movement (PDM), a coalition of 11 political parties led by the PML-N, appears to lack the power to address the issue of missing persons in Balochistan with their subordinate intelligence agencies. For instance, when asked about the rise of missing persons in Balochistan during his first visit to the province in 2022, the PM Shabaz Sharif, stated that he would raise the issue of missing persons in the province with "powerful quarters" (i.e., the Pakistani military and intelligence agencies) (Shah, 2022a).

After assuming the command of Southern Command Quetta in August 2022, Lieutenant General Asif Ghafoor took serious note of the surge in missing persons in line with the Army's policy of WHAM in Balochistan. He is adamant about continuing the Army's policy of WHAM in the province (Qadeer 2022, Interview, 08 December). General Asif Ghafoor said while addressing the youth of Balochistan, "My heart beats with the missing persons and their families .... I'll make sure that if there is evidence [of supporting and abetting insurgency] against someone, they are detained through the proper legal procedure". He assured full support to the families of the missing persons (The Express Tribune Magazine, 2022).

Amnesty and Disarmament, Demobilization, and Reintegration (DDR) Programs

The political objectives of the Balochistan campaign revolve around providing safety to the CPEC via DDR programs and reconciliation opportunities for the insurgents. The SCW doctrine (2013, p. 9) also emphasises the reconciliation, reintegration, and rehabilitation of the insurgents. The Pakistani Army runs the DDR program in Balochistan in collaboration with the federal and provincial governments. Though the amnesty scheme was followed in the earlier round of insurgency, it was not accompanied by pragmatic and concordant reintegration programs. Before 2016, the number of insurgents who surrendered was less than the ones who surrendered post-2016. One of the contributing factors to this was the aggressive practices of the fighting units, killing of insurgents who surrendered during operations. Asad (2019, Interview, 02 January) maintains that over the period this tendency in the fighting troops has considerably been reduced. Thus, the number of insurgents who surrendered to the government has improved considerably. The surrender of the insurgents to the government has been covered, not only by the local press but also by foreign media such as the Hindustan Times (2017), UK Reuters (Yousafzai, 2017), and Xinhua News Agency (2018b), China. Figure 6.3 (below) clearly shows that there is a significant increase in the number of insurgents who surrendered to the security forces. Although there is a difference of figures in these data, their common feature is the upward trajectory of the number of insurgents who surrendered from 2014 onwards. Moreover, these data also show the consistency of the surrender. Even after the change in the provincial and federal political setups in 2021 and 2022 respectively, the consistency of insurgent surrender is not hampered. This is because several programs introduced by the federal and provincial governments for the reintegration of the insurgents in the province continue at the same pace.

The government has initiated several programs for the reintegration of the insurgents and their families in society. These programs are aimed at making them a useful and productive member of society. For example, special classes of various vocational training courses, like plumbing, blacksmithing, electrician work and needlecraft and thread work in different institutes have been arranged for former insurgents and their families by the Balochistan Technical Education and Vocational Training Authority (B TEVTA). Special budget allocation is made every year to the B TEVTA for this purpose. After successfully completing the courses, these former insurgents are given a job as per the technical skill in various government/semi-government departments and autonomous bodies in the province.

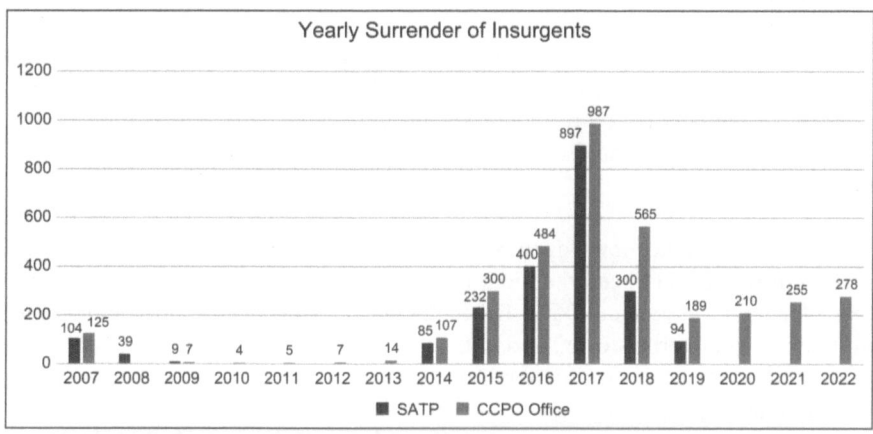

Figure 6.3 Yearly Surrender of Insurgents in Balochistan (2007–2022).[18]

Sources:

a) South Asia Terrorism Portal (SATP). Available at https://www.satp.org/datasheet-terrorist-attack/surrender/pakistan-balochistan (Accessed on 15th January 2023)

b) Capital Chief Police Officer (CCPO) – Quetta, Balochistan. Data collected by the researcher.

Self-employment and entrepreneurship in small and micro enterprises is another option availed by many of these former insurgents (Ahsan 2019, Interview, 16 January). As per the statistics maintained by the CCPO office Quetta, the total number of insurgents who had been reintegrated into the society, till January 2022, after completing vocational training was 2666 men apart from 365 women and 675 juveniles who were the family members of some of these insurgents. In February 2022, there were around 340 men still enrolled in various courses. The former insurgents who completed their vocational training through B TEVTA make up to around seventy-nine per cent of those who surrendered and were granted amnesty, while another ten per cent were still enrolled in various courses in February 2022[19]. This demonstrates the success of the DDR in reducing insurgency in Balochistan.

Governance Extension and Corruption

Poor governance in Balochistan has always been a critical issue. Unfortunately, people suffer in the province due to the corruption of politicians and business executives. For example, Pakistan signed the historic millennium declaration at the United Nations Millennium Summit in 2000. The eight Millennium Development Goals (MDGs) were set in the summit to be achieved by 2015. These goals ranged from halving extreme poverty rates to halting the spread of HIV/AIDS, providing universal primary education and avoiding child and maternal mortality (MGD Fund, n.d.). Twenty-one targets were set at the national level in Pakistan to achieve eight MDGs. Balochistan's performance was worst of all the provinces as it could not achieve a single target (Hassan & Malik, 2018, p. 5). This speaks of the state of governance in the province. Corruption was rampant in government departments. For example, Aamir opines:

> Balochistan is facing myriad problems ranging from insurgency to sectarian violence and bad governance. In such a situation, one of the less focused issues is the rampant corruption in the province which has rendered the government machinery ineffective.
>
> (2015, para.1)

This fact was also acknowledged by the former PM Imran Khan, who maintained that the province had been the victim of large-scale corruption in the funds allocated for its development (Daily Times, 2018). The incapacity of the government in transparent governance coupled with large scale corruption had a direct bearing in worsening the security situation in the province as it impeded the development works especially in the insurgency affected areas.

The current situation of governance coupled with a sharp decline in corruption at the government departments in the province is far better compared to 2015 (Achakzai, 2019). There are many reasons for it. Foremost is the proactive role of the National Accountability Bureau (NAB) in Balochistan since 2016. For example, in 2016, NAB recovered 730 million rupees in currency notes in a raid at the residence of the finance secretary of the government of Balochistan (Shah Syed, 2016).

The finance secretary accumulated this wealth through illegal means and misappropriation of government funds. In 2018, the NAB recovered the total amount of 488.097 million rupees from various persons involved in financial corruption in the province of Balochistan and this amount reduced to 63.219 million rupees in 2021 (Pakistan National Accountability Bureau (NAB) Annual Report, 2018, p. 34; 2021, p. 157). Sana (2022, Interview, 11 August), a Baloch nationalist from Awaran District and Lecturer in Economics at Lasbela University (Balochistan), maintains that the sharp reduction in the NAB recoveries from 2016 to 2021 can truly be attributed to the sharp decline in corruption in the province over the years, especially at the government departments.

The Government of Balochistan has also introduced many programs, in collaboration with international organisations, as initiatives for better governance in the province. For instance, in October 2018, the Government of Balochistan and the United Nations Office on Drugs and Crime (UNODC) launched the Governance Structure of Rule of Law Roadmap in Balochistan. It is a model of good practice encompassing government ownership, a partnership between the State and society, and promotion of a whole-of-government approach for strengthening the criminal justice system in the province (UNODC, 2018).

Many other programs, in collaboration with international organisations, concerning education, health, and rural development, etc., are still in progress in the province. These initiatives help in strengthening the functioning of the concerned provincial department through the technical expertise of the partner organisations like the World Bank, UNDP, etc., and moreover helps in creating jobs for the local people. The districts on the CPEC route, popular with the insurgents and their supporters, are the prime focus of these programs (Ajmal, 2018, Interview, 17 December; Durrani 2022, Interview, 11 August). The LEAs provide thorough security cover in the field for the implementation of these initiatives. That is how these initiatives help in winning the hearts and minds of the people and gaining the support of the population for the counterinsurgents in Balochistan.

Economic Pillar

The economic component of the 3PCM comprises long term programmes for development assistance across a range of infrastructure development, industrial, and commercial activities. Economic deprivation, along with socio-political injustices, was one of the prime causes of the earlier rounds of insurgency in Balochistan. The SCW doctrine (2013, p.2 9) emphasises the importance of economic development in fighting insurgency by stating that the military operations with political ownership take place in a broader framework of the National Security Strategy for creating a friendly environment for sustained development and bringing about better socio-economic change as one of its aims.

It is a fact that Balochistan had been one of the poorest regions in Pakistan. As per the report of the Institute of Public Policy (2011, cited in Ahmed & Baloch, 2017, p. 7) an average resident of Balochistan lived on less than a dollar a day income, about 90 per cent of the settlements in Balochistan had no access to clean drinking water, and the rural illiteracy rate exceeded 90 per cent in the province. The per capita income in Balochistan is less than half of the country's average. These statistics present a grim picture. After 2013, there has been a gradual improvement on the economic side in the province. With the advent of CPEC in the province, a massive resource and infrastructure uplift and management effort have gone in by the Federal and Provincial Governments as compared to the past though still inadequate to produce a very significant result in the vast Balochistan. For instance, after being sworn in as PM in April 2022, Shehbaz Sharif officially visited the port city of Gwadar in Balochistan, announcing the allocation of over 100 billion rupees for various development projects in the province under the Federal Public Sector Development Programme (FPSDP). This development fund will address the water scarcity and power supply issues in Gwadar and Makran districts (Shah, 2022b). The premier vowed that the federal government would provide all possible support to the provincial government in addressing the problems faced by the people of Balochistan. Such initiatives have undoubtedly resulted in development assistance programs in the fields of education, water resource management, power supply, and agriculture. These programs will be further discussed in detail later in this section.

These initiatives will lead to economic growth in the province. The Federal Government realised this situation even back in 2010, and therefore, the share of Balochistan in the Seventh National Finance Commission (NFC) Award[20] was increased from seven per cent to nine per cent (Sadaat, 2019). This was an unprecedented increase in the flow of finances from the central government to Balochistan. Likewise, under the 18th constitutional amendment (Article 172(3)),[21] the provinces were declared as 50 per cent owners of all the natural resources found within their territories. The implication was that Balochistan province would be able to negotiate better royalty rates, with the federation, as a member of the energy corporations in the oil and gas sector (Khalid, 2010). Out of these two initiatives, only the NFC was implemented, and unfortunately, the 18th constitutional amendment clause on the issue of the ownership of natural resources could not be executed[22]. The latter is vital for the economic uplift of Balochistan and diluting the decade's old problem of economic marginalisation. In fact, one of the cardinal points of the insurgents' theme of an independent Balochistan is based on safeguarding the natural resources of the province. As mentioned in the previous chapters, the insurgents consider the Federal Government of Pakistan as the plunderer of the resources of Balochistan. In the aftermath of the CPEC, the insurgents consider China as a party to the Federal Government of Pakistan in the economic colonisation of the province. This amply highlights the importance of the implementation of the 18th constitutional amendment in true letter and spirit to make the insurgent narrative null and void.

The efforts of the Federal Government of Pakistan are not enough to remove the chronic discord between Balochistan and Islamabad on the issue of economic share, especially in the aftermath of CPEC projects. For example, the Provincial Government of Balochistan, in December 2018 and early 2019, voiced its concern over the share of the province in CPEC projects (Shahid, 2018). The main concern of the cabinet members was that during the last five years, only two projects – Gwadar Port and HUBCO coal-based power plant – were approved for Balochistan. The then CM vowed to get the due share for Balochistan (Shahid, 2018).

Sardar Nasir highlights the sensitivity of the issue of disparity in the economic share for Balochistan in the following words:

> The problem of due economic share for Balochistan is no less than the matter of loyalty to the province and its inhabitants and the issue of Baloch nationalism. No political party in Balochistan, even the collation partners of the Federal Government, can compromise the issue of due economic share for the province. If some party dares to make compromises on this sensitive issue so surely, they will be wiped out by people (voters) form the province in the next general elections, and their politics will meet death forever.
>
> (2019, Interview, 10 January)

"Khushal Balochistan" Programme (Prosperous Balochistan)

The Pakistani Military, along with other LEAs, with the assistance of the Federal and Provincial Governments, launched the "Khushal Balochistan" programme in November 2017 which is still in progress (Pakistan Today, 2017). This initiative is aimed at stabilising of Balochistan through socio-economic development and enhanced security measures (ISPR, 2017d) and thus providing the people of Balochistan with civic facilities and a peaceful environment (Pakistan Today, 2018b). Mahmood (2020) captures the soul of this socioeconomic development program in the following words:

> Socioeconomic development, the most important manoeuvre, is the hallmark of Khushal Balochistan. Pakistan Army has remained engaged [through the Khushal Balochistan programme] in numerous development and socioeconomic uplift projects so as to regain the lost space for further capitalization by the political prong, thus bringing lasting peace and stability in the province.

The projects under this programme are specifically tailored for the needs and development of the insurgency ridden areas of the province (ISPR, 2018a) and are mainly humanitarian and development assistance projects. For example, in March 2018, the Frontier Works Organisation (FWO), a subsidiary of the Pakistani Army, with the support of the Swiss and UAE government, established a water desalination plant for the people of Gwadar (ISPR, 2018b). The initial capacity of the plant is 4.4 million gallons of water per day, which could be enhanced to 8.8 million gallons per day. This has ensured the provision of clean drinking water to the local population. Apart from this plant, the FWO has provided technical expertise and logistics support to the Balochistan Public Health Engineering Department for constructing other water supply schemes in the far-flung regions of the province.

The Pakistani Army, in collaboration with the Provincial Government, has established many educational and vocational training institutions in the province while giving priority to the Baloch dominated districts. In 2018, the Pakistani Army's Corps of Engineers started the construction of a cadet college at Jaho, Awaran, the most troublesome district of Balochistan. Cadet College Awaran has been designed initially for 800 cadets, with hostel facilities for 300 cadets and faculty members, with plans to enhance its capacity subsequently (ISPR, 2018c). It is a massive project and near completion. Apart from these initiatives, numerous other initiatives are carried out by the Army and other LEAs on a regular basis. These include free medical camps (The Nation, 2019), flood relief operations (Sasoli, 2019; The Express Tribune, 2022), the establishment of dispensaries, construction, and maintenance of the buildings of many primary and high schools by the FC (Shahid, 2015) and the FWO and many more. Karamat aptly summarises the development initiative of the army in Balochistan in the following words:

> The military has undertaken massive development projects in Balochistan. …… Major institutions have been established for the uplift of the province. The Army Institute of Mineralogy, the Balochistan Institute of Marine Sciences and the Balochistan Institute of Technical Education are functional as are a number of Army Public Schools, vocational training centres, a degree-awarding college, military college, the Gwadar Institute of Technology and the Quetta Institute of Medical Sciences. Five major hospitals have been set up, and eight area development projects have been undertaken by the military across Balochistan. In addition, at least eight infrastructure projects have been completed [by FWO], thereby vastly increasing connectivity. These include roads, tunnels, airfields, a major new international airport at Gwadar Port, a coastal highway and a dam. The army has been involved in water supply, sanitation and irrigation projects, as well as a hydro-power station and two windmill operated electricity-generating projects. The military has also worked to provide sports facilities and encourage local culture and sports activities [in Balochistan].
>
> (2019, p. 81)

Overall, the proactive role of the army in the humanitarian and development assistance programmes launched under the auspices of the Khushal Balochistan initiative has produced positive perceptions, especially in the Baloch dominated and insurgency affected areas of Balochistan. Qasim (2022, Interview, 03 January), who is a local journalist, argues that the Khushal Balochistan initiative has been very beneficial for the local population which is mostly comprised of underprivileged class. For example, in April 2018, 250 free laptops were given to the needy students of Khuzdar district (a Baloch dominated area and heavily affected by insurgency) as part of the "Khushal Balochistan" Program for research and development purposes (The Balochistan Times, 2018). Most of the recipients belonged to the Baloch ethnicity.

The humanitarian and development assistance programmes launched under the auspices of the Khushal Balochistan initiative have helped the military in reinforcing the Psy Ops themes in their favour and exposing the insurgent's propaganda against the military. Adnan (2019, Interview, 14 January) maintains that these humanitarian assistance efforts by the military have even motivated many insurgents to lay down arms voluntarily. The statistics, as shown in Figure 6.3, substantiate this claim. All these efforts are the result of realisation at the army level that assistance to the civil authorities in gearing up the development process in the province is vital. The former COAS, General Bajwa reiterated "Balochistan's progress is Pakistan's progress. Pakistan Army will extend full support and assist the socio-economic development of Balochistan .... All measures in coordination with the civil government would be taken to bring peace and prosperity in Balochistan" (ISPR, 2018b).

Development Programmes by the International Organisations in Balochistan

There are certain economic development programmes being run in the province which are fully or partially funded by international donor bodies. Mostly these pertain to development assistance and resource management. Two of the most notable programs in this regard are the Balochistan Integrated Water Resource Management and Development Project and the Balochistan Education (BE) project, both funded by the World Bank.

Sustainable water management is an important issue for the Government of Balochistan, and the World Bank has commitments of more than $200 million in Balochistan to develop this critical resource for its people (The World Bank, n.d.) by initiating the Water Resource Management and Development Project. The project aims to strengthen provincial government capacity for water resources monitoring and management and to improve community-based water management for targeted irrigation schemes in Balochistan (The World Bank, n.d.). Unfortunately, this project has been suspended mid-way by the World Bank because of lack of progress in managing the project, disbursing funds, proceeding with the civil works, and fiduciary control (The World Bank, 2019). Apart from this, the World Bank has commitments of more than $250 million in Balochistan for implementing a wider programme of support for the province. The principal investments of the World Bank in the province are in education (BE project), health, governance, and water (The World Bank, 2019).

The results of the implementation of the BE project from 2015 to 2018 have been positive (The World Bank, 2018). More than 900 schools in Balochistan are fully functional. Out of these 700 schools are with new/renovated building structures, and 100 schools have been upgraded from primary to middle and high. Under this fund 53,000 students (72 per cent girls) were enrolled with a retention rate of 89 per cent. Almost 700 schools have a comprehensive Early Childhood Education (ECE) Program with trained teachers and ECE specific learning material (The World Bank, 2018). Over 2000 community members in the near vicinity of the schools have been trained to monitor the school construction and teacher presence in schools (The World Bank, 2018). One of the interesting facts about this project was that its focus was on the Baloch dominated areas where the insurgency is prevalent. The police, FC, and the community protection teams provide security to these schools. There has not been a single incident of violence recorded against this project (CCPO Office, 2019; personal communication). Asif (2018, Interview,

07 December) highlights that in the BE project schools, a large number of children who study here belong to the same villages and towns as that of the insurgents.

The economic pillar requires further deliberation by the Federal and Provincial Governments to frame the whole-of-government response strategy. There are many grey areas such as the distribution of resources and non-implementation of the 18th amendment, which resulted in the economic discontent of the people. Moreover, the institutional capacity of the various provincial departments requires enhancement so that the international donor bodies like the World Bank do not suspend their development projects in the province on the issue of lack of governance and corruption. These steps if sincerely implemented, will facilitate sustainable development in the province, enhance the economic growth, develop provincial-federal harmony, and reduce poverty levels in the province. Thus, it will result in gaining maximum population support for the government, and the insurgents' agenda will be defeated. This may help in providing sustainable security to CPEC projects.

Continuation of Army's Policy in Balochistan

The Pakistani Army is committed to the policy of Winning Hearts and Minds (WHAM) in Balochistan even after the change of political government and military leadership in the country in 2022, says Qadeer (2022, Interview, 08 December). The appointment of Lieutenant General Asif Ghafoor as commander of Southern Command Quetta in August 2022 indicates the continuation of the Army's mindset that the military cannot address the root causes of insurgency in Balochistan through kinetic means alone; its job is to facilitate the political process and economic growth by controlling the violence by insurgents and assist the civil authorities in gearing up the development process in the province. This was the approach developed and initiated by General Kiyani and carried forward by his successors. Asif Ghafoor has a vast experience in planning and conducting sub-conventional operations. Asif Ghafoor believes in a non-kinetic approach in the lead with minimum essential unavoidable kinetic part to shape the environment for non-kinetic prongs because he considers that using military force alone could not bring lasting peace in the restive province (Qadeer 2022, Interview, 08 December). "The issue of Balochistan could only be resolved through an inclusive national political process", he said (The Express Tribune Magazine, 2022). "For this purpose, the annoyed Baloch leaders must join the political mainstream", this approach is well aligned with the SCW doctrine and renewed counterinsurgency approach of the Pakistani Army.

Pakistani COIN in Balochistan (post-2013) in the Light of the *Good and Bad COIN Practices Scorecard/Table*

As mentioned in the introductory chapter, Kilcullen's 3PCM does not provide any standard measure for gauging the effectiveness of each pillar, so the *Good and Bad COIN practices scorecard/table* (attached as Appendix 6) developed by Paul et al. (2013a, p. 249) is used for the security pillar. The COIN scorecard is applied purely based upon the qualitative narrative details, collected during field research work in

Pakistan after the significant shift in the overall COIN approach of the Pakistani Army in the light of the SCW doctrine (2013) as it marked the new phase[23] of the fifth round. Finally, to highlight the effectiveness of the COIN practices (security pillar) of the Pakistani Army in Balochistan, the sum of good-minus-bad COIN practices is considered.

Out of the list of the fifteen good COIN practices, the Pakistani COIN force followed six in total (post-2013) as a result of the change in the COIN approach. These are:

a The Pakistani COIN forces *realised at least two strategic communication factors*. One is that the messages or themes of IOs are cohered with the overall COIN approach. All the themes and messages of the information strategy are ultimately based on the political aim of the counterinsurgency campaign in Balochistan, that is, safeguarding the CPEC projects from all threats (Global Times, 2022; Siddiqui, 2019; Xinhua, 2018a; Khan Abdul Wajid, 2018). Secondly, the themes and messages are coordinated across all relevant (involved) government agencies primarily through ISPR and HQ SC.

b The Pakistani COIN force *realised at least one intelligence factor*. As a result of improved real-time actionable intelligence gathering and sharing procedures, as mentioned earlier in the chapter, the capability of the COIN forces was enhanced, facilitating the disruption of insurgent processes and operations.

c The *unity of effort and unity of command* principles are thoroughly practised by the Pakistani COIN forces in Balochistan, as highlighted earlier.

d The Pakistani COIN forces in Balochistan *avoided excessive collateral damage and disproportionate use of force* in their revised strategy.

e The Pakistani COIN force *sought to engage and establish positive relations with the population in the area of conflict*. For example, the manifestation of the concept of "every soldier is an ambassador of the Pakistani Military's goodwill for Baloch" is now practised by the infantry patrols in the conflict-affected areas in the form of informal greeting exchanges and honouring the Baloch tribal culture and values. Moreover, the Pakistani Army resorts to face-to-face communication in the conflict areas with the tribal elders in the form of routine meetings/ Jirgas, dinner, or lunches arranged on various religious and cultural festivals for disseminating the information/themes, as discussed earlier.

f Lastly, *investments, improvements in infrastructure or development, occurred in the area of conflict* controlled or claimed by the Pakistani COIN force. The development projects carried out by the Pakistani Army, like establishing the cadet colleges and schools, water purification schemes (para 6.5.1). Moreover, the 448-kilometre-long N-85 Surab-Hoshab road has been completed at a cost of approximately 22 billion rupees by the FWO. It links Gwadar Port with the National Highway network (N-25) near Surab, Quetta, and other parts of the country (Dunya News TV, 2016). Likewise, the M-8 road project of 200.6 km (Nalient-Turbat-Hoshab) is a major step towards the utilization of the Gwadar port. M-8 is almost complete and links the major city of Balochistan i.e., Quetta and further to Afghanistan through Chamman (FWO, 2017).

As against the six good COIN Practices, the Pakistani COIN force still deploys two bad COIN practices, including the fact that the *fighting was initiated primarily by the insurgents*. The fifth round of insurgency was initiated by the Baloch secessionist insurgents. For example, in 2004 Baloch Liberation Front (BLF) re-emerged in the news after killing three Chinese foreign workers working on a Pakistani mega-development project in Balochistan. In 2005, the BLF carried out multiple attacks, targeting Pakistani Security Forces, pipelines, and foreign workers (Stanford University, 2015a). The Pakistani Army launched the operations in 2006 after successive attacks by the Baloch insurgents especially once the BLA fired rocket hit the helicopter of Inspector General of Frontier Corps (IGFC) in December 2005. Major General Shujaat Zamir Dar (IGFC) was inspecting the Kohlu area (Balochistan) while flying where two days earlier BLA fired eight rockets once the then president General Musharraf was visiting the town (Gulf News, 2005). The other bad COIN practice by the Pakistani COIN force is the *failure to adapt to changes in adversary strategy, operations, or tactics*. This holds true once it comes to the insurgent nexus with the International Jihadists organisation like ISKP in Balochistan where the latter further collaborated with the sectarian militant organisations to disrupt CPEC because of its anti-Chinese and anti-Pakistan agenda, as highlighted in Chapter 5. The Pakistani COIN force is yet to devise some specific strategy, which is of immense importance, to break these nexuses.

This study finds that the sum of good-minus-bad COIN practices of the Pakistani Army in Balochistan is a positive value which depicts the effectiveness of the COIN strategy at large.

One of the very essential COIN aspects which Pakistani forces have realised very late and till now it could not be manifested in Balochistan is the *disruption of tangible support to the insurgents*. Disruption of tangible support to the insurgents and adaptability on the part of COIN forces are two of the specific factors realised by all the successful COIN campaigns completed between WWII and 2010 as per the research of Paul et al. (Paul & Clarke, 2016, p. ix) and it is part of the COIN scorecard/table (Paul et al., 2013a, p. 249). Pakistan could not reduce the flow of cross-border insurgent support because of the porous border with Afghanistan and Iran. These porous borders with the rugged mountainous and desert terrain facilitate the illegal cross-border movement of insurgents from Balochistan to Afghanistan (Lieven, 2017, p. 179) and in some cases to Iran as well. Militants of the Islamic State of Khorasan (ISKP) also use the porous border between Balochistan and Afghanistan for the movement and planning attacks in Pakistan. The issue of the porous border enables insurgents to replenish resources, material (ammunition and weapons) accusation, recruiting, and training across the border and financing. A lack of disruption of tangible support to the insurgents owing to the porous border contributes significantly to the existing violence in Balochistan and thus poses a direct threat to the CPEC. Asad (2019, Interview, 02 January), a battalion commander in Balochistan, magnifies this grave issue by boldly stating that the porous border with Afghanistan and Iran is the prime reasons that the Pakistani Security Forces are unable to end, though it has been greatly reduced, the menace of insurgency and violence from Balochistan as yet.

The scorecard does not guarantee its ability to predict the outcome of the current Balochistan insurgency. The positive value of good-minus-bad COIN practices of the Pakistani Army in Balochistan only offers a good indication that the counterinsurgency efforts of the Pakistani Army are on a path likely to lead to a favourable outcome if the essential factor of *disruption of tangible support to the insurgents* is practically addressed. Moreover, the Pakistani COIN forces need to devise some specific strategy to counter the secessionist insurgent nexus with the ISKP in Balochistan. This is as essential as the disruption of tangible support to the insurgents.

Conclusion

The institutionalisation of the learning followed by its practical manifestation, in the light of SCW doctrine (2013), was a significant milestone achieved by the Pakistani Army after a long journey. The COIN strategy in Balochistan after 2016 is largely a consequence of the doctrinal teachings based on the vast counterinsurgency experience of the army. Safety of the CPEC, particularly in Balochistan, has proved to be a game-changer that has revitalised the entire mindset, organisational culture, and COIN practice of the army.

The Federal Government set the "safety of the CPEC projects from all threats" as the political aim, as well as the national security objective, of the counterinsurgency campaign in Balochistan. Thereafter, the Pakistani Military developed a broad information base. On this firm base, security pillar, political pillar, and economic pillar were developed in parallel. Such a comprehensive COIN strategy is something not known in the counterinsurgency history of the Pakistani Army.

Though these pillars developed simultaneously, they are not in balance, which is again essential for a successful COIN. Out of all these pillars, the security pillar appears to be the firmest one, followed by the political and lastly the economic pillar. The security pillar does not address the issue of global jihadists (ISKP) and their nexus with the sectarian militants and the insurgents in the province of Balochistan. The military component of the security pillar mostly addresses the Baloch secessionist insurgents whereas in the prevailing scenario ISKP and the sectarian militants also pose a significant threat to the security of the CPEC as discussed in Chapter 5. Moreover, it has failed to reduce the tangible support to the insurgents from across the borders. The political pillar suffers from the weak institutional capacity of the civil government and lack of good governance. Finally, the economic pillar is the most disjointed one mainly because of the lack of political consensus on the economic distribution of resources, royalty of the CPEC projects in Balochistan, etc. Despite all these grey areas, the current security, political, and economic situation in Balochistan is far better than ever, which can be interpreted as a source of optimism.

Undoubtedly, there exists a large room for improvement in the COIN strategy in Balochistan. The final chapter "Conclusion" will highlight the cumulative effect of the security, political, and economic strategy implemented in Balochistan at large, as discussed in this chapter, by superimposing the "COIN Scorecard" based on the security-related data sets. Moreover, recommendations for the remedial measures will be posited regarding the weakness in the overall existing COIN strategy.

Notes

1 In contrast, the insurgents rely on spreading their message through pirate FM radio stations in districts like Gwadar and Turbat (Qasim 2022, Interview, 03 January).

2 For example, the exaggerated figures of the killings of the Pakistani Army personnel deployed on the security of CPEC routes/infrastructure (Qasim (a local journalist), 2022, Interview, 03 January; Asim (army officer), 2019, Interview, 25 January).

3 See Bruce Hoffman, "Today's Highly Educated Terrorists", *National Interest*, September 15, 2010; Alan B. Krueger and Jitka Malecková, "Education, Poverty and Terrorism: Is There a Causal Connection?" *Journal of Economic Perspectives*, Vol. 17, No. 4, Fall 2003, pp. 119–144; "Exploding Misconceptions: Alleviating Poverty May Not Reduce Terrorism but Could Make It Less Effective", *Economist*, December 16, 2010.

4 "Coordinated inauthentic behaviour" is when groups of pages and people by creating networks of fake accounts mislead others about who they were and what they were doing. Coordinated inauthentic behaviour is stopped by taking down these pages and accounts because as per the Facebook official policy, they do not want their services to be used to manipulate people" (Gleicher, 2018).

5 A concept unleashed by the former COAS, General Qamar Bajwa during his first visit to the HQ Southern Command Quetta in January 2017 (Ahsan 2019, Interview, 16 January).

6 Hub Power Company Limited, commonly known as HUBCO, is Pakistan's biggest independent power producer. Its power generation is spread across the country with four power generation units already contributing to the national grid.

7 Operation Radd-ul-Fasaad was initially aimed at eliminating any threat of insurgency, militancy, and terrorism from Balochistan and later it was extended throughout the country. Operation Radd-ul-Fasaad will be discussed in detail under the main heading of *Security Pillar of 3PCM*.

8 In Balochistan, 360,000 people suffered from heavy rains and Monsoon flooding in 2022, 238 people were killed, 106 injured, and some 20,000 people were reported to be displaced. More than 500,000 livestock died in Balochistan province, where livestock is a critical source of sustenance and livelihood for many families. At least 17,500 houses have been destroyed, and another 43,900 houses partially damaged. In addition to houses and croplands, 1,000 km of roads and 18 bridges were also damaged and impeded access across flood-affected areas (reliefweb, 2022).

9 The actual number of the recruits, year-wise breakdown, is not known as it was not shared with the researcher in interviews of the military personnel during fieldwork in Pakistan.

10 For example, see Shah, Abid Hussain (2007), The Volatile Situation of Balochistan – Options to Bring It into Streamline. Available at: http://citeseerx.ist.psu.edu/viewdoc/download?doi=10.1.1.466.6333&rep=rep1&type=pdf

11 ISR: An activity that synchronizes and integrates the planning and operation of sensors, assets, and processing, exploitation, and dissemination systems in direct support of current and future operations. This is an integrated intelligence and operations function. (Report of the Defense Science Board Task Force on Defense Intelligence: *Counterinsurgency (COIN) Intelligence, Surveillance, and Reconnaissance (ISR) Operations.* Available at https://apps.dtic.mil/dtic/tr/fulltext/u2/a543575.pdf)

12 The National Action Plan (NAP) was formulated and passed by the Parliament on 24th of December, 2014 as a consequence of a national consensus to curb the militants after the Army Public School, Peshawar (KPK) attack by terrorists on 16th of December 2014 killing more than 148 people and most of them were the children (Hashim, 2018b). Earlier in February during the same year the National Internal Security Policy (NISP) was announced by the Federal Government. It was the first ever document in the civilian domain to address the need for an integrated security approach (Weinbaum, 2017; p. 49). After NISP, NAP was the second document in a row passed by the parliament. A 20-point NAP included the lifting the moratorium on executions, the establishment

of special military courts for militants and terrorists and clamping down the religious extremism and provisions for regulating madrassas throughout the country (Weinbaum, 2017, p. 49). The NAP also envisaged an active role of the moribund National Counter Terrorism Authority (NACTA) enacted in 2009. "However, most of the objectives of NAP have been piecemeal and not pursued with any great determination" (Weinbaum, 2017, p. 49). As far as the military strategy was concerned, there was absolutely no change after NAP; the military continued *Operation Zarb-e-Azb* in North Waziristan (FATA) with massive force, and in the same way, heavy-handed military operations were conducted in Balochistan as well.

13 In Baloch language, Parari refers to an individual or group with genuine grievances that cannot be resolved through dialogue [For details, see Harrison, 1978, p. 30].

14 Reportedly, the Pakistani army acquired the strike-capable CH-4B drone, capable of carrying a 345 kg weapon payload with flight endurance of 14 hours.

15 Data for 2022 is not available at UCDP website.

16 The design of the fence is two parallel fences 3-4 meters apart are erected. The fence stands 13 feet high on the Afghan side and 11 feet on the Pakistani side, sharp spirals of silver barbed wire are cradled at the top of each. Additional coils of Dannert Wire layered over one another have been placed on the ground in the gap between the two rows of fences, which are dotted with Pakistani Military towers (Yusufzai, et al., 2018).

17 In this regard, SCW doctrine (2013) is no different than any other counterinsurgency manual/doctrinal document of some professional army, see for example, the U.S. Army/ Marine Corps Counterinsurgency Field Manual 2007: 3–24.

18 South Asia Terrorism Portal (SATP) has no data beyond 2019.

19 The former insurgents report to their respective territory's police station on fortnightly basis for two years and once a month during the third year after being granted amnesty. Moreover, they have to inform the police in case they move out of the city and report to the police station on return as well. All the police stations of the province furnish the data regarding the appearance before the police and movement, of the former insurgents, to the CCPO office Quetta as a monthly report. These are the check and balance procedures in vogue in Balochistan for the former insurgents. That is why the data exists with the CCPO office (Official at CCPO office Quetta, 2022, personal communication). The B TEVTA officials refused to share any information about insurgent vocational training courses with the researcher in line with the official policy of the department. Moreover, the exact employment figures of the former insurgents could not be traced because neither the HQ SC nor the CCPO office Quetta had any consolidated statistics. The concerned section officer of the Civil Secretariat (Balochistan Government) in Quetta did not share the figures regarding the employment pattern of the insurgents with the researcher.

20 Pakistan is a federation and has four federating units: Sindh, Punjab, Khyber-Pakhtunkhwa (KPK), and Balochistan, as well as other territories including Gilgit-Baltistan and Azad Jammu and Kashmir which come under the federal government directly. The NFC award is meant to distribute financial resources between the federal government (vertical distribution), and the provinces (horizontal distribution) (Adeney, 2012; Tariq, 2019).

21 Unanimously adopted by parliament in 2010 after two years of deliberations by a parliamentary committee that was represented by all the major parties, the 18th Amendment virtually overhauled the 1973 Constitution. The amendment includes 102 important articles and has made the 1973 Constitution more democratic (Hussain, 2019).

22 For example, Mian Raza Rabbani (2018), the Chairman Senate of Pakistan, observed in a ruling on 'Operationalization of joint ownership of mineral oil and natural gas; implementation of Article 172 (3), Constitution, 1973',

> ……. it is clear that no serious measure/ step has so far been taken by the Federal Government to implement the Article 172(3) in letter and spirit. Minor consultation with the Provinces on selective matters shall not be construed as implementation of the Article 172(3) …… Implementation of Article 172(3) requires

complete over haul of the existing law, administration, management and regulatory regime related to mineral oil & natural gas.

(Pakistan Senate, 2018, p. 7)

23 The shift in the COIN approach of the Pakistani Army in Balochistan after promulgating the SCW doctrine (2013) marks the initiation of a new phase of the fifth round of COIN in the province. Paul et al., set the phase demarcation criteria as follows, "A new phase was declared when the case analyst recognized a significant shift in the COIN approach, in the approach of the insurgents, or in the overall conditions of the case" (2013a, p. 16).

References

Aamir, A. (2015) 'War Against Economic Terrorism in Balochistan', *The Balochistan Point,* 13 September. Available at: http://thebalochistanpoint.com/war-against-economic-terrorism-in-balochistan/ (Accessed: 28 August 2022).

Achakzai, J. (2019) 'Plugging into Balochistan and its governance through conflicting geopolitical interests', *The News*, 23 November. Available at: https://www.thenews.com.pk/print/572884-plugging-into-balochistan-and-its-governance-through-conflicting-geopolitical-interests (Accessed: 15 December 2022).

Adeney, K. (2012). 'A Step towards Inclusive Federalism in Pakistan? The Politics of the 18th Amendment', *Publius. The Journal of Federalism*, 42 (4), (Fall 2012), pp. 539–565. doi: 10.1093/publius/pjr055.

Ahmed, M. & Baloch, A. (2017). 'The Political Economy of Development: A Critical Assessment of Balochistan, Pakistan'. *Munich Personal RePEc Archive (MPRA)*, Paper No. 80754. Available at: https://mpra.ub.uni-muenchen.de/80754/1/MPRA_paper_80754.pdf (Accessed: 21 July 2022).

Ali, I. & Stewart, P. (2018) 'Exclusive – As Trump cracks down on Pakistan, U.S. cuts military training programmes', *Reuters,* 11 August. Available at: https://www.reuters.com/article/pakistan-usa-military-idINKBN1KW07X (Accessed: 11 August 2022).

Alvi, M. (2018) 'About 300 missing persons return home in Balochistan: BNP-M', *The News*, 19 October. Available at: https://www.thenews.com.pk/print/382782-about-300-missing-persons-return-home-in-balochistan-bnp-m (Accessed: 19 August 2022).

ANI (2023) *In 2022, Pak army killed 195, forcibly disappeared 629 people in Balochistan, says Paank Annual Report 2022*. Available at: https://www.aninews.in/news/world/asia/in-2022-pak-army-killed-195-forcibly-disappeared-629-people-in-balochistan-says-paank-annual-report-202220230112170854/ (Accessed: 20 April 2023).

Arab News (2022) *Pakistan army confirms general commanding southern 12 Corps killed in helicopter crash*. Available at: https://www.arabnews.pk/node/2134386/pakistan (Accessed: 20 April 2023).

Associated Press of Pakistan (APP) (2017) *Balochi youth as capable as youth of other areas of country: COAS*. Available at: https://www.app.com.pk/balochistan-engine-of-future-development-effort-coas/ (Accessed: 26 July 2022).

Aziz Khalid. (2010) 'Important Features of 7th NFC Award and 18th Amendment'. *The Pakistan Development Review*, 49 (4) (Winter 2010), pp. 537–542. Available at: https://www.jstor.org/stable/41428674?seq=1#metadata_info_tab_contents (Accessed: 21 July 2022).

Baloch Human Rights Council (2022) *Two hundred one went missing, and 59 persons extrajudicially killed in Balochistan*. Available at: https://balochhumanrightscouncil.wordpress.com/2022/07/17/two-hundred-one-went-missing-and-59-persons-extrajudicially-killed-in-balochistan/ (Accessed: 20 April 2023).

Baloch Human Rights Council (2023) *Balochistan: The State of Human Rights in 2022*. Available at: https://hrcbalochistan.com/balochistan-the-state-of-human-rights-in-2022/ (Accessed: 20 April 2023).

Balochistan Levies Force. (2021). *About Us*. Available at: https://balochistanleviesforce. gob.pk/about/ (Accessed: 20 April 2023).

Basit, S. H. (2019). 'Terrorizing the CPEC: Managing Transnational Militancy in China–Pakistan Relations', *The Pacific Review*, 32 (4), pp. 694–724. doi: 10.1080/09512748.2018.1516694.

BBC News (2022) *Pakistan attack: China condemns killing of tutors in Pakistan blast*. Available at: https://www.bbc.com/news/world-asia-61225678 (Accessed: 20 April 2023).

Business Recorder (2022) *Sheikh Sultan donates $2.5m to Army Relief Fund: ISPR*. Available at: https://www.brecorder.com/news/40199517 (Accessed: 20 April 2023).

Business Standard (2019) *Plane with 'World Must Speak Up for Balochistan' banner flies over Edgbaston stadium*. Available at: https://www.business-standard.com/article/news-ani/plane-with-world-must-speak-up-for-balochistan-banner-flies-over-edgbaston-stadium-119071101241_1.html (Accessed: 17 August 2022).

Clark, D. J. (2006). *The Vital Role of intelligence in Counterinsurgency Operations*. Carlisle Barracks, PA: U.S. Army War College.

CPEC Authority. (2022). *Energy Projects Under CPEC*. Available at: https://cpec.gov.pk/energy (Accessed: 27 March 2022).

Daily Quetta Voice (2022) *Balochistan Assembly pays rich tribute to Shaheed Lt. General Sarfaraz*. Available at: https://www.quettavoice.com/2022/08/06/balochistan-assembly-pays-rich-tribute-to-shaheed-lt-general-sarfaraz/ (Accessed: 20 April 2023).

Daily Times (2018) *Bad governance, corruption behind Balochistan's backwardness: PM*. Available at: https://dailytimes.com.pk/333060/bad-governance-corruption-behind-balochistans-backwardness-pm/ (Accessed: 17 June 2022).

Davidovic, J. (2018). 'Displacement as Significant Collateral Harm in War', *Global Justice: Theory Practice Rhetoric (TPR)*, 11 (1), pp. 64–84. doi: 10.21248/gjn.11.1.136.

Dawn (2019) *Balochistan govt approves special security division for Quetta*. Available at: https://www.dawn.com/news/1479896 (Accessed: 17 August 2022).

Dunya News TV (2016) *PM Nawaz to inaugurate N-85 Surab-Hoshab road today*. Available at: http://dunyanews.tv/en/Pakistan/365303-PM-Nawaz-to-inaugurate-N85-SurabHoshab-road-toda (Accessed: 13 June 2022).

FWO (2017) *Gawadar – Turbat – Hoshab (M-8)*. Available at: https://fwo.com.pk/projects/ongoing-projects/highways/407-gawadar-turbat-hoshab-m-8 (Accessed: 27 March 2022).

Galula, D. (1964) *Counterinsurgency Warfare: Theory and Practice*. Reprint, Westport, CT: Praeger, 2006.

Gleicher, N. (2018) *Coordinated Inauthentic Behavior Explained*, 06 December. Available at: https://newsroom.fb.com/news/2018/12/inside-feed-coordinated-inauthentic-behavior/ (Accessed: 21 August 2022).

Gleicher, N. (2019) *Removing Coordinated Inauthentic Behavior and Spam from India and Pakistan*, 01 April. Available at: https://newsroom.fb.com/news/2019/04/cib-and-spam-from-india-pakistan/ (Accessed: 17 August 2022).

Global Times (2022) *Exclusive full interview with Pakistani ambassador: China-Pakistan ties are significant regardless of leadership change*. Available at: https://www.globaltimes.cn/page/202204/1259660.shtml (Accessed: 20 April 2023).

Gul, A. (2018) 'Pakistan's Fencing of Afghan Border Remains Source of Mutual Tensions', *Voice of America (VOA)*, 15 October. Available at: https://www.voanews.com/a/pakistan-s-fencing-of-afghan-border-remains-source-of-mutual-tensions/4614787.html#:~:text=Pakistan%27s%20unilateral%20installation%20of%20a,a%20busy%20southwestern%20crossing%20point. (Accessed: 8 July 2022).

Gulf News (2005) *Attack on helicopter wounds army commander*. Available at: https://gulfnews.com/world/asia/pakistan/attack-on-helicopter-wounds-army-commander-1.308193 (Accessed: 03 June 2022).

Guramani, N. (2017) 'COAS briefs senators on national security', *Dawn*, 19 December. Available at: https://www.dawn.com/news/1377559 (Accessed: 10 August 2022).

Harrison, S. (1978). 'Nightmare in Baluchistan', *Foreign Policy*, 32 (Autumn), pp. 136–160.

Hashim, A. (2018b) 'In Pakistan, wounds of Peshawar school attack reopen each winter', *AlJazeera,* 16 December. Available at: https://www.aljazeera.com/news/2018/12/pakistan-wounds-peshawar-school-attack-reopen-winter-181216074137219.html (Accessed: 19 August 2022).

Hassan, M. & Malik, U. M. (2018) *Development Advocate Pakistan: Balochistan Challenges & Opportunities*. Islamabad: UNDP. Available at: https://www.undp.org/pakistan/publications/balochistan-challenges-opportunities (Accessed: 21 August 2022).

Hindustan Times (2017) *Around 500 Baloch rebel militants surrender, pledge allegiance to Pakistan.* Available at: https://www.hindustantimes.com/world-news/around-500-baloch-rebel-militants-surrender-pledge-allegiance-to-pakistan/story-qbk6xnZtd2Z0sHaYW-cR6jJ.html (Accessed: 15 August 2022).

Hoffman, B. (2004). *Insurgency and Counterinsurgency in Iraq.* Santa Monica, CA: RAND Corporation. Available at: https://www.rand.org/content/dam/rand/pubs/occasional_papers/2005/RAND_OP127.pdf (Accessed: 25 February 2022).

Hussain, Z. (2019) 'Debating 18th Amendment', *Dawn*, 06 February. Available at: https://www.dawn.com/news/1462145 (Accessed: 14 August 2022).

Iqbal, K. (2018). 'Securing CPEC: Challenges, Responses and Outcomes', in Arduino, A. & Gong, X. (eds.) *Securing the Belt and Road Initiative: Risk Assessment, Private Security and Special Insurances along the New Wave of Chinese Outbound Investments.* Singapore: Palgrave Macmillan, pp. 197–214.

ISPR. (2017a). *Press Release No PR-87/2017.* 22 February. Available at: https://www.ispr.gov.pk/press-release-detail.php?id=3775 (Accessed: 19 June 2022).

ISPR. (2017b). *Press Release No PR-88/2017.* 23 February. Available at: https://www.ispr.gov.pk/press-release-detail.php?id=3776 (Accessed: 19 June 2022).

ISPR. (2017c). *Press Release No PR-159/2017.* 25 March. Available at: https://www.ispr.gov.pk/press-release-detail.php?id=3864 (Accessed: 11 June 2022).

ISPR. (2017d). *Press Release No PR-591/2017.* 05 December. Available at: https://www.ispr.gov.pk/press-release-detail.php?id=4422 (Accessed: 19 July 2022).

ISPR. (2018a). *Press Release No PR-101/2018.* 02 March. Available at: https://www.ispr.gov.pk/press-release-detail.php?id=4606 (Accessed: 19 August 2022).

ISPR. (2018b). *Press Release No PR-102/2018.* 04 March. Available at: https://www.ispr.gov.pk/press-release-detail.php?id=4607 (Accessed: 17 July 2022).

ISPR. (2018c). *Press Release No PR-104/2018.* 05 March. Available at: https://www.ispr.gov.pk/press-release-detail.php?id=4609 (Accessed: 19 August 2022).

Jaffrelot, C. (2014) *Coming Out of the Cold*, Carnegie Endowment for International Peace. Available at: https://carnegieendowment.org/publications/57613 (Accessed: 22 October 2022).

Janes (2021) *Pakistan receives five CH-4 UAVs from China.* Available at: https://www.janes.com/defence-news/news-detail/pakistan-receives-five-ch-4-uavs-from-china (Accessed: 17 March 2022).

Karamat, J. (2019). 'Sustained Political Progress: The Supportive Role of the Military', in Burki, S. J., Chowdhury, I. A. & Butt, A. E. (eds.) *Pakistan at Seventy: A Handbook on Developments in Economics, Politics and Society.* Abingdon, Oxon: Routledge, pp. 73–84.

Kermani, S. & Mudassir, M. (2017). 'Balochistan journalists caught 'between the stick and the gun', *BBC News,* 26 November. Available at: https://www.bbc.co.uk/news/world-asia-42086695 (Accessed: 11 July 2022).

Khan Abdul Wajid. (2018) 'Threats against CPEC, Pakistan-China ties', *China Daily,* 29 November. Available at: http://www.chinadaily.com.cn/a/201811/29/WS5bffa6e6a-310eff30328be00.html (Accessed: 11 June 2022).

Khan, A. (2009). 'Renewed Ethnonationalist Insurgency in Balochistan, Pakistan: The Militarized State and Continuing Economic Deprivation', *Asian Survey; Berkeley,* 49 (6), pp. 1071–1091. doi: 10.1525/as.2009.49.6.1071.

Raza, K. (2016) '15,000 troops of Special Security Division to protect CPEC projects, Chinese nationals', *Dawn,* 12 August. Available at: https://www.dawn.com/news/1277182 (Accessed: 11 August 2022).

Khan Wajahat, S. (2015) 'Ex-Balochistan Militants Recount Paths to War with Pakistan', *NBC News,* 30 August. Available at: https://www.nbcnews.com/news/world/former-balochistan-militants-recount-path-war-pakistan-n417036 (Accessed: 17 August 2022).

Khyber News (2019) *Balochistan's apex committee reviews implementation of NAP.* Available at: https://khybernews.tv/balochistans-apex-committee-reviews-implementation-of-nap/ (Accessed: 17 August 2022).

Kilcullen, D. (2006b) *'Three Pillars of Counterinsurgency'.* Available at: http://www.au.af. mil/au/awc/awcgate/uscoin/3pillars_of_counterinsurgency.pdf; https://tamilnation.org/armed_conflict/060928kilcullen.htm (Accessed: 22 January 2018).

Lambert, L. (2019) 'Brawling Afghanistan and Pakistan fans face World Cup bans as ICC investigate how a plane trailing a political message evaded air traffic control to fly over Headingley', *Mail Online,* 30 June. Available at: https://www.dailymail.co.uk/sport/cricket/article-7198425/Brawling-Afghanistan-Pakistan-fans-face-World-Cup-bans-ICC-investigate-plane-fly-over.html (Accessed: 17 August 2022).

Lieven, A. (2017). 'Counter-Insurgency in Pakistan: The Role of Legitimacy', *Small Wars & Insurgencies,* 28 (1), pp. 166–190. doi: 10.1080/09592318.2016.1266128.

Mahmood, S. (2020) 'Balochistan: A Journey Towards Prosperity', *Hilal.* Available at: https://www.hilal.gov.pk/eng-article/detail/NDM5OQ==.html (Accessed: 19 August 2022).

Mangi, F. (2023) 'Why Pakistan Is Struggling to Get Another IMF Bailout', *The Washington Post,* 06 February. Available at: https://www.washingtonpost.com/business/energy/why-pakistan-is-struggling-to-get-another-imf-bailout/2023/02/04/3d446f86-a508-11ed-8b47-9863fda8e494_story.html (Accessed: 20 April 2023).

MDG Fund (n.d.). *Millennium Development Goals.* Available at: http://www.mdgfund.org/node/922 (Accessed: 11 December 2022).

Meer, S. (2015) 'CPEC: A Bad Deal for the Baloch People?', *The Diplomat,* 30 December. Available at: https://thediplomat.com/2015/12/cpec-a-bad-deal-for-the-baloch-people/ (Accessed: 21 July 2022).

Menezes, J. (2019) 'Pakistan and Afghanistan cricket fans involved in fight outside Headingley after plane flew over ground with "Justice for Balochistan" message', *Independent,* 29 June. Available at: https://www.independent.co.uk/sport/cricket/pakistan-afghanistan-fans-fight-video-cricket-world-cup-outside-ground-plane-justice-for-balochistan-a8980421.html (Accessed: 17 August 2022).

Naseemullah, A. (2014). 'Police Capacity and Insurgency in Pakistan', in C. Christine Fair, and Sumit Ganguly (ed.) *Policing Insurgencies: Cops as Counterinsurgents.* Oxford International Relations in South Asia, pp. 177–202

Naseer, G. K. (2010). *Tareekh-e-Balochistan (History of Balochistan)* (5th ed.). Quetta: Kalat Publisher.

National Accountability Bureau (NAB). (2018). *Annual Report-2018.* Available at: http://syklibrary.ndu.edu.pk/libmax/opac/BookDetail.aspx?id=82014 (Accessed: 11 June 2019).

National Accountability Bureau (NAB). (2021). *Annual Report-2021*. Available at: https://nab.gov.pk/Downloads/NAB_Annual_Report_2021.pdf (Accessed: 21 August 2022).

Pakistan Senate. (2018). *Ruling of the Chair: Operationalization of Joint Ownership of Mineral Oil and Natural Gas; Implementation of Article 172 (3), Constitution*, 1973. Available at: https://senate.gov.pk/uploads/documents/1582527957_947.pdf (Accessed: 21 June 2022).

Pakistan Today (2017) *COAS announces launch of Khushhal Balochistan project.* Available at: https://archive.pakistantoday.com.pk/2017/11/23/coas-announces-launch-of-khushhal-balochistan-project/ (Accessed: 27 July 2022).

Pakistan Today (2018a) *Balochistan assembly passes resolution to demand due share in CPEC*. Available at: https://archive.pakistantoday.com.pk/2018/12/22/balochistan-assembly-passes-resolution-to-demand-due-share-in-cpec/ (Accessed: 27 March 2022).

Pakistan Today (2018b) *Corps commanders pledge to continue fight for peace, stability*. Available at: https://archive.pakistantoday.com.pk/2018/03/06/corps-commanders-conference-vow-to-continue-fight-for-peace-stability/ (Accessed: 11 June 2022).

Pakistani Army. (2013). *Sub Conventional Warfare (SCW) Doctrine (Publication No. AP 2601E)*. Rawalpindi, Pakistan: Doctrine & Evaluation Directorate, GHQ. Available at: https://www.scribd.com/document/474069737/Sub-Conventional-Warfare-Doctrine-pdf (10 August 2022)

Paul, C. & Clarke, C. P. (2016). *Counterinsurgency Scorecard Update: Afghanistan in Early 2015 Relative to Insurgencies since World War II*. Santa Monica, CA: National Defence Research Institute, RAND Corporation.

Paul, C., Clarke, C. P., Grill, B. & Dunigam, M. (2013a). *Paths to Victory, Lessons from Modern Insurgencies*. Santa Monica, CA: National Defence Research Institute, RAND Corporation.

Paul, C., Clarke, C. P., Grill, B. & Dunigam, M. (2013b). *Paths to Victory, Detailed Insurgency Case Studies*. Santa Monica, CA: National Defence Research Institute, RAND Corporation.

Qureshi, Z. (2019) 'Pakistan to fence 950 km of border with Iran', *Gulf News*, 23 February. Available at: https://gulfnews.com/world/asia/pakistan/pakistan-to-fence-950km-of-border-with-iran-1.62257071 (Accessed: 11 July 2022).

Radio Pakistan (2019) *Complete implementation on National Action Plan necessary: Imran Khan*. Available at: http://www.radio.gov.pk/21-04-2019/complete-implementation-on-national-action-plan-necessary-imran-khan (Accessed: 19 August 2022).

Radio Pakistan (2022) *PM praises Pakistan Army, officials for evacuation of people from Kumrat*. Available at: https://www.radio.gov.pk/28-08-2022/pm-praises-pakistan-army-officials-for-evacuation-of-people-from-kumrat (Accessed: 20 April 2023).

Reliefweb (2022) *Rapid Needs Assessment Report – 2022 Monsoon Floods – Balochistan, Pakistan (August 2022)*. Available at: https://reliefweb.int/report/pakistan/rapid-needs-assessment-report-2022-monsoon-floods-balochistan-pakistan-august-2022 (Accessed: 20 April 2023).

Sadaat, M. (2019) 'The question of governance in Balochistan', *The Friday Times,* 10 May. Available at: https://thefridaytimes.com/10-May-2019/the-question-of-governance-in-balochistan (Accessed: 29 August 2022).

Safi, M. & Baloch, S.M. (2019) 'Armed militants storm luxury hotel in Pakistan attack', *The Guardian*, 11 May. Available at: https://www.theguardian.com/world/2019/may/11/armed-militants-storm-luxury-hotel-in-gwadar-pakistan (Accessed: 16 August 2022).

Sameed, H. (2017). 'Press in Balochistan', *The Express Tribune,* 23 November. Available at: https://tribune.com.pk/story/1565771/6-press-in-balochistan/ (Accessed: 27 July 2022).

Sasoli, I. (2019) '1,500 families rescued from Balochistan's flood-hit areas with army's help: ISPR', *Dawn,* 03 March. Available at: https://www.dawn.com/news/1467413 (Accessed: 25 March 2022).

Shah, A., S. (2022a) 'Missing persons: PM to broach issue with powerful quarters', *The Express Tribune*, 23 April. Available at: https://tribune.com.pk/story/2353929/missing-persons-pm-to-broach-issue-with-powerful-quarters (Accessed: 20 April 2023).

Shah, A., S. (2022b) 'PM Shehbaz announces Rs. 100b development package for Balochistan', *The Express Tribune*, 24 June. Available at: https://tribune.com.pk/story/2363137/pm-shehbaz-announces-rs100b-development-package-for-balochistan (Accessed: 20 April 2023).

Shah Kriti, M. (2017) 'Radd-ul-Fasaad assessing Pakistan's new counter-terrorism operation', *Observer Research Foundation (ORF)*, 01 April. Available at: https://www.orfonline.org/research/radd-ul-fasaad-assessing-pakistans-new-counterterrorism-operation/ (Accessed: 1 September 2022).

Shah Syed, A. (2016) 'NAB recovers Rs. 730m from Balochistan finance secretary's residence', *Dawn*, 06 May. Available at: https://www.dawn.com/news/1256686 (Accessed: 19 August 2022).

Shah Syed, A. (2019) '200 Baloch missing persons have returned home so far this year: home minister', *Dawn,* 29 June. Available at: https://www.dawn.com/news/1491126 (Accessed: 19 August 2022).

Shahid, S. (2015) 'Joint efforts have brought Balochistan out of crisis, say army, govt', *Dawn,* 22 August. Available at: https://www.dawn.com/news/1202096 (Accessed: 11 March 2022).

Shahid, S. (2018) 'Balochistan voices concern over its share in CPEC projects', *Dawn,* 10 December. Available at: https://www.dawn.com/news/1450491/balochistan-voices-concern-over-its-share-in-cpec-projects (Accessed: 26 August 2022).

Shahwani, S. U. (2023) 'Provincial Apex Committee Balochistan Takes Important Decisions', *Discover Baluchistan*, 08 February. Available at: https://www.discoverbaluchistan.com/2023/02/provincical-apex-committee-balochistan.html (Accessed: 20 April 2023).

Shams, S. (2016) 'How Pakistan's military uses media for image making', *Deutsche Welle (DW)*, 16 June. Available at: https://www.dw.com/en/how-pakistans-military-uses-media-for-image-making/a-19335614 (Accessed: 19 June 2022).

Siddiqui, N. (2019) 'Pakistan, China vow to safeguard CPEC from all threats in "strategic dialogue" between foreign ministers', *Dawn*, 19 March. Available at: https://www.dawn.com/news/1470602 (Accessed: 26 July 2022).

South China Morning Post (2015) *Uygur militants 'eliminated' from Pakistan, claims minister*. Available at: https://www.scmp.com/news/china/diplomacy-defence/article/1869252/pakistan-says-it-has-eliminated-uygur-militants (Accessed: 17 August 2022).

Stanford University (2015a) *Mapping Militant Organizations: Balochistan Republican Army (BRA)*. Available at: http://web.stanford.edu/group/mappingmilitants/cgi-bin/groups/view/571?highlight=baloch#cite26 (Accessed: 14 July 2022).

Tariq, W. (2019) 'Explainer: What is the NFC award?', *The Express Tribune*, 25 March. Available at: https://tribune.com.pk/story/1936990/1-explainer-nfc-award/ (Accessed: 27 August 2022).

The Balochistan Post (2019) *10 years of disappearance – A story of three Baloch Missing*. Available at: https://www.youtube.com/watch?v=2aoAEI6DjYc (Accessed: 18 June 2022).

The Balochistan Post (2023) *Baloch population scattered around the world*. Available at: https://thebalochistanpost.net/2023/02/baloch-population-scattered-around-the-world/ (Accessed: 20 April 2023).

The Balochistan Times (2018) *250 students in Khuzdar awarded laptops through "Khushal Balochistan" Program.* Available at: https://dailybalochistantimes.com/quetta/%ef%bb%bf250-students-in-khuzdar-awarded-laptops-through-khushal-balochistan-program/ (Accessed: 19 December 2022).

The Economic Times (2022) *Pakistan Army says defence budget for 2022–23 decreases from 2.8% of the GDP to 2.2%.* Available at: https://economictimes.indiatimes.com/news/defence/pakistan-army-says-defence-budget-for-2022-23-decreases-from-2-8-of-the-gdp-to-2-2/articleshow/92213022.cms?from=mdr (Accessed: 20 April 2023).

The EurAsian Times (2022) *Chinese CH-4B Drone Spotted Over Balochistan; Reports Indicate Pakistan Army Is Using Them to Crush Rebellion.* Available at: https://eurasiantimes.com/chinese-ch-4b-drone-spotted-over-balochistan-reports-indicate/ (Accessed: 20 April 2023).

The Express Tribune (2022) *Army continues flood relief operations across Pakistan.* Available at: https://tribune.com.pk/story/2372482/army-continues-flood-relief-operations-across-pakistan (Accessed: 20 April 2023).

The Express Tribune Magazine (2022) *Balochistan Needs a Healing Touch.* Available at: https://tribune.com.pk/story/2388327/balochistan-needs-a-healing-touch (Accessed: 20 April 2023).

The Frontier Post (2018) *200 anti-state militants lay down weapons in Balochistan.* Available at: https://thefrontierpost.com/200-anti-state-militants-lay-weapons-balochistan/ (Accessed: 17 August 2022).

The Nation (2018) *Education in Balochistan.* Available at: https://nation.com.pk/28-Aug-2018/education-in-balochistan (Accessed: 26 July 2022).

The Nation (2019) *Pak Army establishes free medical camp in Balochistan.* Available at: https://nation.com.pk/04-May-2019/pak-army-establishes-free-medical-camp-in-balochistan (Accessed: 10 August 2022).

The Tribune (2019) *Pak Army to raise another division to protect CPEC projects, Chinese nationals.* Available at: https://www.tribuneindia.com/news/world/pak-army-to-raise-another-division-to-protect-cpec-projects-chinese-nationals/775348.html (Accessed: 11 August 2022).

The World Bank (2018) *Pakistan: Addressing Poverty and Conflict through Education in Balochistan.* Available at: https://www.worldbank.org/en/results/2018/10/25/pakistan-addressing-poverty-conflict-education-balochistan (Accessed: 12 August 2022).

The World Bank (2019) *World Bank Statement on Balochistan Integrated Water Resource Management and Development Project.* Available at: https://www.worldbank.org/en/news/press-release/2019/03/28/world-bank-statement-on-balochistan-integrated-water-resource-management-and-development-project (Accessed: 28 August 2022).

The World Bank (n.d.) *PK-Balochistan Integrated Water Resources Management & Development Project.* Available at: http://projects.worldbank.org/P154255?lang=en (Accessed: 13 July 2022).

Thompson, R. (1966). *Defeating Communist Insurgency.* Reprint, St. Petersburg, FL: Hailer Publishing, 2005.

United Nations Office on Drugs and Crime (UNODC). (2018). *A Significant Milestone Achieved towards Implementation of the Rule of Law Roadmap in Balochistan.* Available at: https://www.unodc.org/pakistan/en/the-government-of-balochistan-and-undoc-launch-governance-structure-of-rule-of-law-roadmap-in-balochistan.html (Accessed: 21 June 2022).

US Department of the Army. (2005). *Psychological Operations* (Field Manual 3-05.30, Marine Corps Reference Publication 3-40.6). Available at: https://fas.org/irp/doddir/army/fm3-05-30.pdf (Accessed: 17 August 2022).

Weinbaum, M. G. (2017). 'Insurgency and Violent Extremism in Pakistan', *Small Wars & Insurgencies*, 28 (1), pp. 34–56. doi: 10.1080/09592318.2016.1266130.

Xinhua (2018a) *Full text of China-Pakistan joint statement*. Available at: http://www.xinhuanet.com/english/2018-11/04/c_137581441.htm (Accessed: 11 July 2022).

Xinhua (2018b) *265 militants surrender in Pakistan's Balochistan province*. Available at: http://www.xinhuanet.com/english/2018-09/19/c_137477924.htm (Accessed: 11 July 2022).

Yousafzai, G. (2017) 'Pakistan says over 300 Baloch separatist militants surrender', *Reuters (UK),* 10 December. Available at: https://www.reuters.com/article/uk-pakistan-militants/pakistan-says-over-300-baloch-separatist-militants-surrender-idUKKBN1E40L6 (Accessed: 10 August 2022).

Yusufzai, M., Whittaker, F., Khan Wajahat, S. & Mengli, A. (2018) 'Pakistan is building a fence along border with Afghanistan', *NBC News*, 17 May. Available at: https://www.nbcnews.com/news/world/pakistan-building-fence-along-border-afghanistan-n873291 (Accessed: 19 July 2022).

Zafar, M. (2017) 'COAS calls for harnessing Balochistan's human resource', *The Express Tribune,* 07 December. Available at: https://tribune.com.pk/story/1578643/1-army-chief-says-believes-democracy-selfless-service/ (Accessed: 26 July 2022).

Zubair, M. (2019) 'Balochistan Deserves Justice in Chinese Projects', *Gandhara*, Available at: https://gandhara.rferl.org/a/pakistan-balochistan-deserves-justice-in-chinese-projects-cpec/29708813.html (Accessed: 13 June 2022).

7   Conclusion

This chapter has two main parts, the essential findings/conclusions and policy recommendations on a prescriptive note. After highlighting the book's rationale, the chapter begins by synthesising the main arguments of the study. Then it highlights the key takeaways that emerge from the study. Thereafter, the chapter proffers some recommendations for renewed counterinsurgency (COIN) policymaking in the local and global context for China/Pakistan in achieving their strategic goals of Belt and Road Initiative (BRI)/China Pakistan Economic Corridor (CPEC). Lastly, it explains how the innovative conceptual modelling of the study and rich empirical research can be utilised for understanding insurgencies in other regions in Asia where BRI will be extended or implemented soon or in the long run.

The principal objective of this book is to determine the utility of the Pakistani Army's COIN strategy after enacting the Sub-Conventional Warfare (SCW) doctrine in 2013, during the fifth round of insurgency in Balochistan for achieving the safety of the CPEC; BRI.

This study concludes, articulated through the empirical analysis presented in the previous chapters, that the transition in the COIN strategy of the Pakistani Army in the light of the Sub-Conventional Warfare (SCW) doctrine (2013) has significantly reduced the collateral damage and fatalities as a result of counterinsurgency activities in Balochistan till 2019. This renewed COIN strategy has also resulted in the increased number of insurgents surrendering, which has led to a considerable drop in insurgent activities like IEDs, landmines, and bomb/hand grenade blasts in Balochistan province. The renewed COIN strategy in the light of the SCW doctrine (2013) has thus helped in managing the security situation in Balochistan which is a good sign for the smooth progress of the CPEC in the province.

Chapter 1 sets out the intellectual foundations of the book's central premise, namely, how has the Pakistani Army's COIN Strategy (2013–2022) contributed to reducing the perceived threats to the CPEC in Balochistan? It also traces the evolution and broad contours of the Balochistan Insurgency in its historical context and the ongoing fifth round of the Balochistan insurgency's correlation with the security of the CPEC. In particular, it cast the debate surrounding the ongoing fifth round of Balochistan insurgency and the security of the CPEC. In short, the first chapter introduced the main scope of the book by crystallising the research questions and the main argument.

DOI: 10.4324/9781003413905-7

Chapter 2 elaborates the methodology and conceptual framework of the study. This study has benefited from utilising Downie's institutional learning cycle to identify the doctrinal development stages of the Pakistani Army. Kilcullen's (2006) Three Pillars of Counterinsurgency Model (3PCM) was also applied to examine the effects of the counterinsurgency strategy of the Pakistani Army in Balochistan, mainly the Information Operations (IOs), security, political, and economic components, after the promulgation of the SCW doctrine in 2013. This framework enabled the evaluation of a "whole-of-government" counterinsurgency strategy keeping in view the background and situation of Balochistan.

Chapter 3 sets out the development of the Pakistani Army's counterinsurgency doctrine in a historical context. It traces the Pakistani Army's culture of fighting *small wars*. Then it looks at the step-by-step development of the institutional learning of the Pakistani Army towards fighting insurgency, using Downie's Institutional Learning Cycle. The chapter also highlights the Lessons Learned during various counterinsurgency campaigns/periods, which depict the Pakistani Army's flexibility and willingness to learn.

Chapter 4 analyses the institutionalisation of the learning of the Pakistani Army concerning counterinsurgency. It explains the institutionalisation process of the learning initiated by General Kiyani (former COAS) and the efforts and stimuli to reach the consensus of promulgating the SCW doctrine. The chapter argues that the institutionalisation learning of the Pakistani Army ultimately resulted in the *Change in Organisational Behaviour;* the changed mindset from a conventional war focus to the acceptance of the fact that COIN demands a different approach than "butcher and bolt" and specialised training in the light of SCW doctrine (2013). Thereafter, the chapter looks at the themes enshrined in the SCW doctrine and how they address the performance gaps of the army highlighted in Chapter 3.

Chapter 5 deals with the threat matrix of the CPEC. It focuses on analysing and evaluating multiple external threats vis a vis the prevailing geopolitical security situation in Balochistan and adjoining areas of Afghanistan and Iran. The chapter finds that apart from external threats, there are diverse internal threats to CPEC. The biggest internal threat for CPEC is the ongoing secessionist insurgency in Balochistan, followed by the activities of sectarian militant organisations in the province. Other threats include the Islamic State of Khorasan (ISKP) in Balochistan and its alliance formation with various insurgent and sectarian militant organisations in the province. This scenario increases the hostile environment for CPEC in Balochistan. The chapter finds that Balochistan's existing volatile security environment, directly challenges the SCW doctrine of the Pakistani Army on which the army based its operations in Balochistan.

Chapter 6 deals with characterizing the Pakistani Army's role, assistance, approach, and campaigns in Balochistan using Kilcullen's three pillars of the counterinsurgency model from 2006 to 2012 (prior to the counterinsurgency doctrine) and 2013 to 2022 (after formally enacting the counterinsurgency doctrine). The continuity of each pillar and the balance between them, as maintained by the Pakistani Army, is discussed in detail. The role of the military is also highlighted in-depth, which apparently seems part of the security pillar only, but of course military means are applied across the model, not just in the security domain. The chapter argues that the renewed COIN strategy in Balochistan after 2016 is largely a consequence of the doctrinal teachings based on the vast counterinsurgency experience of the army. Safety of the CPEC, particularly in Balochistan, has proved to be a game-changer that has revitalised the army's entire mindset, organisational culture, and COIN practices.

This research is vital because it has direct policy consequences for the Chinese BRI in South Asia in general and particularly for the CPEC. At present, there is no study which is specifically focused on the eradication of insurgency from Balochistan in the context of CPEC. This study has evaluated the measures taken by the Pakistani Army against the ongoing fifth round of the secessionist insurgency in Balochistan for providing security to the CPEC in the province where ISKP has developed alliances with the Baloch separatist groups and the sectarian militant organisations.

Key Findings and Research Conclusions

Key findings emerge from this study through critical analysis presented in the preceding chapters. One of the findings is that certain structural factors impede the learning process of counterinsurgency. Large scale interstate war is the predominant factor which hampers the learning process of counterinsurgency and linked to it is the strong bias of the army towards conventional warfighting. Other structural factors which impede learning include the geostrategic priorities of the military. These all hold true in the case of the Pakistani Army's institutional learning process of counterinsurgency. There were broadly four learning cycles within the Pakistani Army, and only the last one (September 11, 2001–2013) could reach to the final stage and thus result in a change in organisational behaviour with regards to learning SCW. All other learning cycles were broken at some stage and could not proceed to the terminal point. One of the primary reasons was the ever-looming threat of war with India, which materialised in four large scale conventional wars between India and Pakistan in 1948, 1965, 1971, and 1999. Moreover, during this time (1948–2001) the senior military leadership was traditionalist and believed in only conventional warfighting in an India-Pakistan war scenario. This belief was based on factual realities. Conventional wars with India in 1948 and 1965 were the prime reason for breaking the first two out of the three learning cycles which could not progress to the final stage of change in organisational behaviour. Again, during the third institutional learning cycle (1966–2001), the 1971 War, and the Kargil War (1999) between Pakistan and India were one of the reasons, apart from traditionalist Pakistani Military leadership being earlier busy in fomenting Afghan Jihad against Soviets at the behest of CIA, that the institutional learning cycle of counterinsurgency of the Pakistani Army was broken. Interruption in the learning cycle, owing to all of the above factors, led to non-publication of the COIN doctrine before 2013 by the Pakistani Army.

The lessons learned in all the learning cycles of the Pakistani Army were quite similar. The political component – the primacy of the political aim – which is one of the most fundamental aspects of the COIN doctrine was missing in the first three institutional learning cycles of the Pakistani Army. During the fourth learning cycle, the political component was initially missing until General Musharraf stepped down as the COAS.

One of the essential findings of the study is that in counterinsurgency campaigns, poor beginnings do not necessarily lead to poor ends and changing practices can lead to improved outcomes. For instance, the Pakistani Army's initial practices were counterproductive. These included extensive use of force, sole reliance on kinetic actions, the abduction and torture of Baloch people and collective punitive actions against the population in the conflict areas. These resulted in heavy casualties of the civilian population. The armed forces also suffered a substantial loss of life. After conceptualising the SCW doctrine, the COIN approach of the Pakistani Army significantly changed, especially from 2016 onwards once the renewed COIN approach was mainly manifested in the province of Balochistan. The changed COIN approach in light of the SCW doctrine (2013) brought some better results in terms of the political primacy of the COIN campaign, reduction of violence/fatalities, WHAM initiatives, etc. The 2013 SCW doctrine furnishes guidance broadly in three domains which include security and military operations, political, and legal aspects, and finally the Civil-Military Coordination (CIMIC) and winning hearts and minds (WHAM) initiatives. These three dimensions of guidance by the SCW doctrine (2013) correspond to Kilcullen's 3PCM.

The study establishes that there are internal and external threats to CPEC in Balochistan. The attacks by the Baloch insurgent organisations such as Baloch Liberation Army (BLA), Baloch Republican Army (BRA), and the Balochistan Liberation Front (BLF) in recent years have adversely impacted the Chinese investment projects in Balochistan. The dimension of the collaboration of the Baloch insurgent organisations amongst each other, with various sectarian organisations, splinter groups of TTP, and finally ISKP has added lethality to these attacks. The province of Balochistan, because of the prevailing insurgency, the deteriorating law and order situation, its sharing of a porous border with Afghanistan and Iran, the presence of large numbers of Shiite and Zikri communities and above all CPEC (particularly Gwadar Port) provides the troika of ISKP, sectarian militant organisations and insurgent organisations with a good opportunity to collaborate in conducting terrorist activities across the length and breadth of the province. ISKP strategy of establishing effective alliances with the local groups in Balochistan by tapping into their grievances based on converging ideologies (either anti-Pakistani or anti-Chinese) poses a potent security threat to CPEC. The trend of suicide attacks, the common practice of Islamic State, in Balochistan during recent years is the outcome of this collaboration. These attacks have resulted in inhibiting the free movement of Chinese nationals in the region. These alliances are soon likely to transform into enduring networks, considerably more potent, with the ability to move resources, human resources, and logistics across the borders of Pakistan, Afghanistan, and Iran which may extend up to Xinjiang later if not checked effectively.

With the frequency of violent attacks by the BLA and LeJ in collaboration with ISKP on the rise in the province during 2020 and 2021, the Pakistani military has resorted to using force out of frustration among its ranks regarding the relatively new phenomenon of international jihadists (ISKP). The Pakistani military lacks any strategy to check the alliance formed between the local insurgent organisations, such as BLA, and the sectarian militant groups like LeJ and the international Jihadists like ISKP in the region. The SCW (2013) is also silent on disrupting the nexus of local insurgents and militants with global jihadists.

The study concludes: (a) security-wise CPEC is a high-risk project in Balochistan; (b) both Pakistan and China are conscious of the security issues pertaining to CPEC and display high resolve in public to secure the project; and (c) the success of the CPEC depends on the stabilisation of the security situation in Balochistan as a high-security threat to the project lies in the province.

The study finds that the Pakistani Army's COIN strategy, even after the promulgation of the SCW doctrine in 2013, took a considerable time to be manifested on the ground in Balochistan. The institutionalisation of the learning and later followed by a practical manifestation in Balochistan in light of the SCW doctrine (2013) was a prominent landmark in the recent military history of the Pakistani Army after a long journey. The political ownership and legitimacy of the COIN campaign in Balochistan is central to the renewed COIN strategy of the Pakistani Army in Balochistan after enacting the doctrine. The Federal Government of Pakistan set the "safety of the CPEC projects from all threats" as the political aim, as well as the national security objective, of the counterinsurgency campaign in Balochistan. Within this ambit of the political aim and the national security objective, the Pakistani Military developed a multi-pronged strategy incorporating a broad information base and developed security, political, and economic measures in collaboration with the Federal Government (and to an extent with the Provincial Government of Balochistan) for controlling the insurgency.

The purpose of the information campaign was to consolidate the messages into a coherent information strategy for the COIN campaign. It included an array of activities such as Information Operations (IOs), media operations, intelligence collection, and dissemination to various agencies involved in the COIN campaign and various countermeasures to the insurgent ideology, motivation, and propaganda. The three pillars of security, politics, and economic measures have been practically developed in parallel, unlike the earlier rounds of insurgency and the period before 2016. Such a comprehensive COIN strategy is something not known in the counterinsurgency history of the Pakistani Army.

This book has detailed that the Disarmament, Demobilization, and Reintegration (DDR) programs and reconciliation opportunities through amnesty for the insurgents proved to be very useful during the COIN campaigns. The government has initiated several programs for the reintegration of the insurgents and their families in society. These programs are aimed at making them a useful and productive member of the community. In Balochistan, special classes of various vocational training courses, such as plumbing, blacksmithing, electrician work, and needlecraft and thread work in different institutes have been arranged for these former insurgents and their families. After completing the courses, the former insurgents are given jobs, as per their technical skills, in various government/semi-government departments and autonomous bodies in the province. The data (refer Figure 6.3) clearly shows that there is a significant increase in the number of insurgents who surrendered to the security forces over the years. These former insurgents prove to be a great assistance in countering the insurgent ideology and agenda in the eyes of the public in their respective towns through their living examples.

The research finds that the military component of the COIN strategy explicitly showed a transition from the conventional practices to the modern doctrine, and it appears to be the firmest one amongst all the three pillars. Unlike in the past, the Pakistani Army is not currently focused on the kinetic means to end the ongoing round of insurgency in Balochistan. The Pakistani Army has adopted a more balanced approach to COIN in light of the SCW doctrine (2013), which is an outcome of effective institutional learning. As a result, the counterinsurgency strategy of the Pakistani Army in Balochistan has been greatly transformed.

Following the strategy of the Intelligence Based Operations (IBOs) with the code name Operation Radd-ul-Fasaad in the context of enhanced actionable intelligence gathering and timely dissemination has helped in reducing deaths amongst the civilian population. It has also enabled the army to commit a force to the COIN operations in Balochistan which is appropriately scaled, effect-based, and focused in the application, unlike the excessive use of force in the past. This has helped in significantly reducing the fatalities, as a result of the state-based violence in Balochistan, after 2016 till 2019.

The military component of the security pillar mostly addresses the Baloch secessionist insurgents whereas in the prevailing scenario ISKP and the sectarian militants also pose a significant threat to the security of the CPEC. Moreover, it has failed to reduce the tangible support to the insurgents from across the porous borders with Afghanistan and Iran. Furthermore, the book establishes that the role of the Balochistan Police and Balochistan Levies Force in the overall matrix of the Pakistani COIN operations in the province has not been given due deliberation. The SCW doctrine (2013) only cursorily touches upon the role of these two LEAs, to the extent that the word "Levies" appears only once in the 100-page doctrinal document. The political pillar suffers from the lack of institutional capacity, transparency, and good governance of the civil government in Balochistan. Finally, the economic pillar is the most disjointed one because of the lack of political consensus on the economic distribution of resources, royalties of the CPEC projects in Balochistan, etc. Though these pillars developed simultaneously, they are not in balance, which is again essential for a successful counterinsurgency strategy. Out of all these pillars, the security pillar appears to be the strongest one, followed by the political and lastly the economic pillar.

As far as the COIN scorecard/table is concerned this study finds that against the six good COIN practices adopted by the Pakistani Army, the Pakistani COIN force mainly resorts to two bad COIN practices even in the aftermath of the change in the COIN approach after promulgating the SCW doctrine in 2013. This study also finds that the sum of good-minus-bad COIN practices of the Pakistani Army in Balochistan is a positive value which depicts the effectiveness of the COIN strategy at large. Furthermore, in answering the main research question, this book concludes that despite all the grey areas mentioned above, the existing security, political, and economic situation in Balochistan has improved. The counterinsurgent forces have controlled the insurgency to a large extent. At the same time, the COIN strategy in Balochistan requires improvement to bring the insurgency to an end and to ensure the smooth functioning of the CPEC. This entails two underlying aspects as already highlighted, firstly the reduction of tangible support to the insurgents through the porous borders and secondly an effective strategy to break the nexus and collaboration of the Baloch insurgents with the ISKP and the latter's collaboration with the local sectarian militants in Balochistan.

The existing body of academic literature on the security situation in Balochistan is relatively scant, particularly regarding its interplay with the CPEC. Additionally, there is a notable dearth of research on the present wave of insurgency in Balochistan and the corresponding COIN strategy undertaken by the Pakistani Army to secure the CPEC in Balochistan. This book stands as a pioneering work in comprehensively examining the ongoing fifth round of Balochistan insurgency, as well as the efficacy of the Pakistani Army's COIN strategy in securing the CPEC in Balochistan. The book effectively addresses the significant gap in the current research landscape through this analysis.

Policy Recommendations

The study is helpful to the Pakistani Government, which is very keen to implement CPEC smoothly, a unique project of massive investment in the country, which is unprecedented in the region. Moreover, this study is equally important for the Chinese government, which has until now invested over $42 billion in the CPEC and this investment will reach up to $62 billion, as Beijing is very concerned about the security and smooth implementation of the CPEC in Balochistan. This study offers the following six policy recommendations based on its key findings.

1   The Pakistani Army needs to revise and update its SCW doctrine (2013) to make it more comprehensive and in line with the latest prevailing COIN scenarios, especially in Balochistan. It is time that the SCW doctrine (2013) is updated with the post-2013 experience of the fifth round of Balochistan insurgency. As per the guidelines of the D&E Directorate of the Pakistani Army, the doctrine needs updating every five years. This revised version should address the critical issues such as the nexus between the insurgents and the ISKP and between ISKP and the sectarian militant organisations. Moreover, it should also address the growing extremism and religious/sectarian violence in Balochistan. A more proactive and efficient role by the Police and Levies Force, in the overall COIN matrix, needs to be considered for taking practical measures to control the extremism and religious/sectarian violence in Balochistan. Furthermore, the revised doctrine requires the inclusion of additional guidelines on reducing tangible support to the insurgents through the porous borders. Thorough deliberation is needed regarding the role of the Balochistan, Levies Force in the overall mechanism for controlling the porous borders of the province with Afghanistan and Iran. Historically, the Balochistan Levies Force has protected the Af-Pak and Pakistan-Iran borders in Balochistan.

2 The Pakistani Army in financial and technical collaboration with the Chinese government should devise a comprehensive border surveillance mechanism for Afghanistan's and Iran's borders with Balochistan. The first step is the priority fencing of the borders, as done in KPK province, which at the moment has stalled due to financial pressures. This should be followed by the technological surveillance of the borders by using technology like drones, CCTV cameras, etc. The Chinese government should earmark the budget for the border surveillance project in Balochistan along with providing technical expertise and support to the Pakistani forces.

3 Private security companies, from Pakistan, should be incorporated in the overall security apparatus of the CPEC in general and in Balochistan in particular. Although the Pakistani Army raised a Special Security Division (SSD) in 2016 at the cost of $220 million (Basit, 2019, p. 706) this arrangement is inadequate. This is the total force (13,700 personnel) for guarding over 2200 km network of roads, railways, and pipelines, Special Economic Zones, power plants, Gwadar Port, and all the affiliated infrastructure along with over 2000 Chinese nationals working on the CPEC from Gwadar to Khunjerab Pass in the north. The existing security personnel strength of all military and other LEAs in Balochistan is not adequate for the security of the CPEC/Chinese personnel, carrying out IBOs, providing human security and public safety. There are not enough security personnel in the province keeping in mind the 134,050 sq. mi. area (nearly the size of Germany) of the province, constituting 44 per cent of Pakistan's total landmass, the diversity and scattered CPEC projects, the length of the CPEC routes, the on-going insurgency, sectarian militancy and the presence of ISKP. Recruiting the regular troops of the army and FC is a financially cumbersome procedure, and it requires considerable procedural time as well. The deficiency of the troops can be made efficiently and economically through incorporating the private security companies from Pakistan. While employing these private security companies, care should be taken that these companies preferably employ the Balochi and Pashtun workforce and avoid the Punjabi personnel as it may spark the ethnic grievances of the Baloch population in Balochistan in general.

4 The government of Pakistan should take pragmatic steps to bring the religious seminaries (madrassas) in the province of Balochistan under government control to ensure effective checks on the activities and funding of these madrassas. From 2010 onwards, Islamic radicalisation gradually increased and hit its peak in Balochistan, led by a Deobandi madrassa network. After 2010, Balochistan has seen the highest number of enrolments in madrassas in Pakistan along with the rapid growth in the number of madrassas (Fair, 2014, p. 18). The sectarian organisations also recruit manpower from the Baloch dominated districts, mainly through madrassas, based on Deobandi ideology. These madrassas are contributing to the sectarian violence in the province, especially the killing of Hazaras, a Persian speaking Shiite community based in Balochistan.

5 Implementation of the 18th constitutional amendment in true letter and spirit is imperative to address the economic discontent of the people in Balochistan. It will help in countering the insurgent's decades-old narrative of the economic

injustices with Balochistan and the comparatively new narrative of the economic colonisation of the province. Moreover, it will also help in eradicating the reservations of nationalist Baloch political leaders like Akhter Mengal.

6 A consensus should be developed between the Federal Government and the Provincial Government of Balochistan over the economic issues and distribution of resources at NFC, which is the best political forum for this purpose.

Another significant element of the study is that its innovative conceptual modelling and rich empirical research can be utilised for understanding insurgencies in other regions in Asia where BRI will be extended or implemented in the near future or the long run. Some of the countries where China is planning to extend BRI have a history of insurgency, and in some cases, the insurgency is still smouldering in the tribal societies of those countries. For instance, the Chinese government plans to expand BRI to Afghanistan (Bonesh & Devonshire-Ellis, 2023; Stone, 2019), where a jihadist insurgency by Al Qaeda and the Islamic State continues (Bunzel, 2022). Afghanistan has porous borders with Pakistan, Iran, and the Central Asian States.

Likewise, China and Myanmar signed a MoU in September 2018 to formalise the China-Myanmar Economic Corridor (CMEC) as a part of BRI (Banerjee, 2022; Yhome, 2019). Multiple ethnic insurgencies have existed in Myanmar since independence (Cline, 2009, p. 576). The porous borders between Myanmar and neighbouring countries have had a significant impact on the course of the insurgencies (Cline, 2009, p. 585). The ethnic resistance, insurgencies, and the brutal counterinsurgency campaigns of the Myanmar army will continue to pose problems for the region (Cline, 2009, p. 588).

In the same manner, the Nepal-China Trans-Himalayan Multi-Dimensional Connectivity Network (THMCN) is part of China's ambitious BRI (Reuters, 2022; IANS, 2019) formalised through a mutual agreement between the two countries in April 2018 (The Hindu Business Line, 2018). Nepal has a long history of Maoist insurgency. In 2019, the splinter Maoist group, which calls itself the Communist Party of Nepal, engaged in violent activities such as detonating over a hundred improvised explosive devices (IEDs), targeting rallies of prominent politicians across the country. The same year the Nepalese government officially declared the Communist Party of Nepal a criminal group and banned all its activities. While the ban was imposed after an established campaign of violence, it carried the risk of escalating the situation into a full-blown insurgency (Adhikari, 2019) as in the past by the Maoists. Nepal has porous borders as well.

Although the Bangladesh-China-India-Myanmar (BCIM) Economic Corridor is no longer listed under the BRI umbrella (Aneja, 2019; IAS Express, 2022) owing to Indian reservations, it may later form part of the BRI. The BCIM corridor is also planned to pass through the north-eastern region of India if implemented, which is again an insurgency laden area with porous borders.

The findings and recommendations of this study are highly relevant for all the planned economic corridors mentioned above under the BRI rubric. This is due to many similar features, such as the history of insurgency or active insurgency in these regions. Along the economic corridor routes in these regions, the societies are tribal, and these regions have porous borders which already facilitated the existing insurgents. Another common dimension of all these potential corridors with the insurgency in Balochistan is the (potential) presence of ISIS. In Afghanistan, ISIS is already present, and in other regions, it is making headway. For example, in April 2019, the director general of Myanmar's President's Office said that Myanmar had been a target of ISIS which has mainly nurtured home-grown [terrorists] cells by working with radical elements inside the country (Zaw, 2019). In India, ISIS claimed to have established the "Wilayah of Hind", a "province" (Tomlinson, 2019). ISIS has also claimed to gain a strong foothold in the disputed region of Kashmir after clashes between the security forces and the militants in May 2019 (Tomlinson, 2019). In Nepal, there is no known pro-ISIS group, but its porous borders with India and rapidly increasing emigrant population in the Middle East raises concerns about the future presence and activities of ISIS in the country (Pant & Taneja, 2019). Lastly, in most of these countries mentioned above, there is a history of military interference in political decision making as in the case of Pakistan.

The security of the CPEC is of paramount importance, and the counterinsurgency campaign in Balochistan plays a crucial role in ensuring it. The Pakistani Army's conceptualization of the SCW doctrine and its subsequent implementation significantly changed the COIN approach in Balochistan. The inclusion of critical components like political primacy, affect-based and focused use of force, winning "hearts and minds", and rules of engagement resulted in a marked reduction in violence, fatalities and an increased number of insurgent surrenders. However, the reduction of tangible support to the insurgents through porous borders and the need to break the nexus of the ISKP in Balochistan is urgently required to end the insurgency and ensure CPEC's security. There is a dire need to re-assess the security situation in Balochistan by understanding the causal mechanisms that lead to alliance formation, drawing some pertinent conclusions, and then devising a comprehensive strategy at the highest level in the country to deal with it in the context of CPEC. This entails the updation of the SCW doctrine, including the guidelines about disrupting the nexus of local insurgents and sectarian militants with global jihadist organizations.

To conclude, the crucial role of Balochistan in ensuring Pakistan's future stability cannot be overstated. Its strategic location, rich resources, resilient population, and significant infrastructure projects like the CPEC and Gwadar Port are indispensable for Pakistan's progress and prosperity. Therefore, it is imperative for Pakistan's military and political leadership to urgently evaluate the latest security, political, and economic conditions of Balochistan and find a way forward in collaboration with each other.

References

Adhikari, G. (2019) 'The spectre of a new Maoist conflict in Nepal', *Aljazeera*, 21 April. Available at: https://www.aljazeera.com/indepth/opinion/spectre-maoist-conflict-nepal-190404111507371.html (Accessed: 02 September 2022).

Aneja, A. (2019). 'Bangladesh-China-India-Myanmar (BCIM) Economic Corridor no longer listed under BRI umbrella', *The Hindu*, 28 April. Available at: https://www.thehindu.com/news/international/bangladesh-china-india-myanmar-bcim-economic-corridor-no-longer-listed-under-bri-umbrella/article26971613.ece (Accessed: 21 September 2022).

Banerjee, S. (2022). 'Revamping BRI in post-coup Myanmar', *Observer Research Foundation (ORF)*, 08 December. Available at: https://www.orfonline.org/expert-speak/revamping-bri-in-myanmar-post-coup/ (Accessed: 15 March 2023).

Basit, S. H. (2019). 'Terrorizing the CPEC: Managing Transnational Militancy in China–Pakistan relations', *The Pacific Review*, 32 (4), pp. 694–724. doi: 10.1080/09512748.2018.1516694.

Bonesh, F. R. & Devonshire-Ellis, C. (2023). 'The Future of the Belt and Road Initiative in Afghanistan: Obstacles, Opportunities, and The Taliban's Perspective', *Silk Road Briefing*, 20 February. Available at: https://www.silkroadbriefing.com/news/2023/02/20/the-future-of-the-belt-and-road-initiative-in-afghanistan-obstacles-opportunities-and-the-talibans-perspective/ (Accessed: 11 April 2023).

Bunzel, C. (2022). 'Explainer: The Jihadi Threat in 2022', *Wilson Center*, 22 December. Available at: https://www.wilsoncenter.org/article/explainer-jihadi-threat-2022 (Accessed: 17 March 2023).

Cline, L. E. (2009). 'Insurgency in Amber: Ethnic Opposition Groups in Myanmar', *Small Wars & Insurgencies*, 20 (3–4), pp. 574–591. doi: 10.1080/09592310903027215.

Fair, C. C. (2014) 'Does Pakistan Have a Madrasah Problem? Insights from New Data'. *SSRN*, 06 August. Available at: https://papers.ssrn.com/sol3/papers.cfm?abstract_id=2468620 (Accessed: 27 April 2022).

IANS (2019) *China majorly increasing footprint in Nepal: Experts*. Available at: https://www.thequint.com/news/hot-news/china-majorly-increasing-footprint-in-nepal-experts (Accessed: 3 September 2022).

IAS Express (2022) *BCIM Economic Corridor – Need for Revival*. Available at: https://www.iasexpress.net/in-depth-bcim-economic-corridor-need-for-revival/ (Accessed: 07 March 2023).

Kilcullen, D. (2006) *'Three Pillars of Counterinsurgency'*. Available at: https://tamilnation.org/armed_conflict/060928kilcullen.htm (Accessed: 22 January 2022).

Pakistani Army (2013). *Sub Conventional Warfare (SCW) Doctrine (Publication No. AP 2601E)*. Rawalpindi: Doctrine & Evaluation Directorate, GHQ.

Pant, H. V. & Taneja, K. (2019). 'ISIS's New Target: South Asia', *Foreign Policy (FP)*, 02 May. Available at: https://foreignpolicy.com/2019/05/02/isiss-new-target-south-asia/ (Accessed: 9 September 2022).

Reuters (2022) *China, Nepal agree on building a trans-Himalayan network*. Available at: https://www.reuters.com/world/china/china-nepal-agree-building-trans-himalayan-network-2022-08-11/ (Accessed: 21 October 2022).

Stone, R. (2019) 'Slowly but surely, China is moving into Afghanistan', *TRT World*, 18 February. Available at: https://www.trtworld.com/magazine/slowly-but-surely-china-is-moving-into-afghanistan-24276 (Accessed: 21 September 2022).

The Hindu Business Line (2018) *Nepal-China agree to expand connectivity, build railway line*. Available at: https://www.thehindubusinessline.com/news/world/nepal-china-agree-to-expand-connectivity-build-railway-line/article23631018.ece (Accessed: 03 September 2022).

Tomlinson, H. (2019) 'ISIS claims a foothold in India after killing troops', *The Times*, 13 May. Available at: https://www.thetimes.co.uk/article/isis-claims-a-foothold-in-india-after-killing-troops-w6kshz2ww (Accessed: 6 September 2022).

Yhome, K. (2019). 'Emerging dynamics of the China-Myanmar economic corridor', *Observer Research Foundation (ORF)*, 15 May. Available at: https://www.orfonline.org/expert-speak/emerging-dynamics-of-the-china-myanmar-economic-corridor-50847/ (Accessed: 3 September 2022).

Zaw, H. N. (2019) 'ISIS a Threat to N. Rakhine: Gov't Spokesperson', *The Irrawaddy*, 29 April. Available at: https://www.irrawaddy.com/news/burma/isis-threat-n-rakhine-govt-spokesperson.html (Accessed: 19 September 2022).

Appendices

Appendix 1 Map of Pakistan

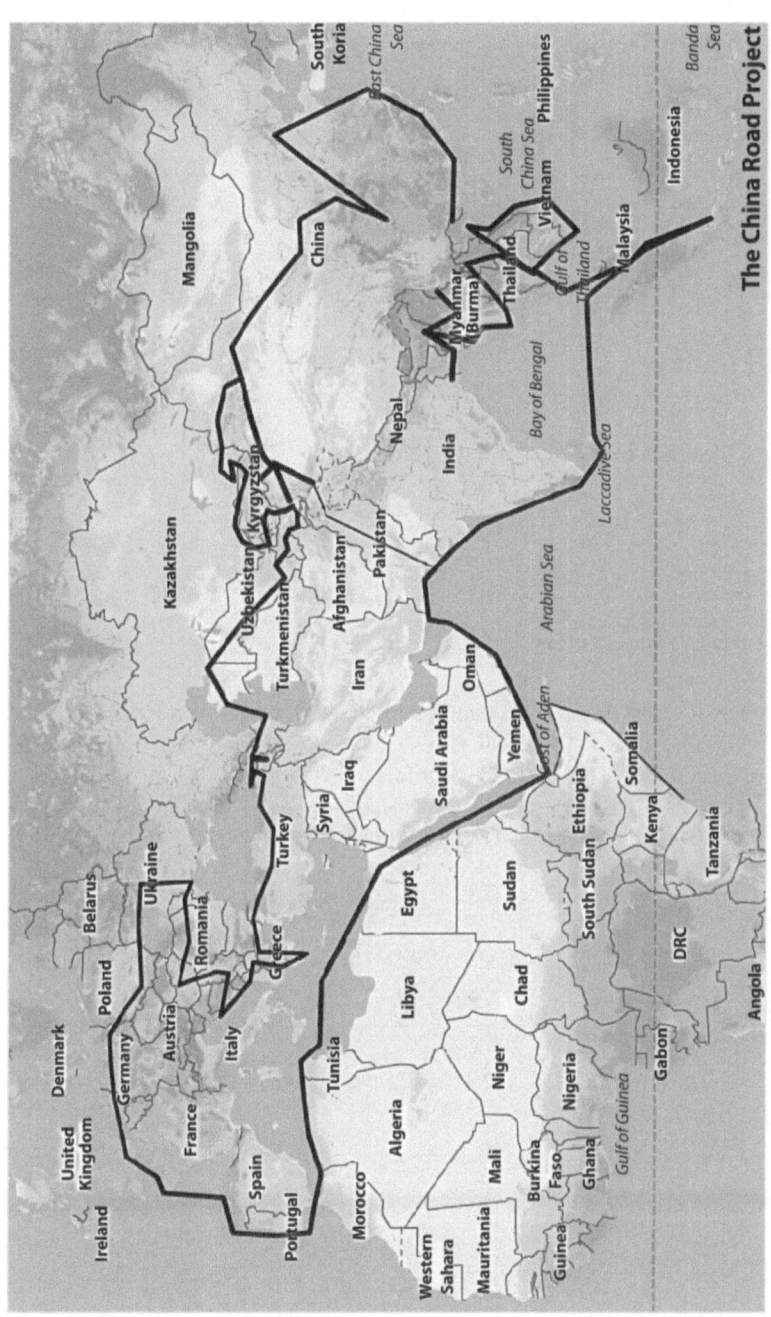

Appendix 2 Map of China's Belt and Road

Source: Reconnecting Asia

Available at: https://reconnectingasia.csis.org/analysis/entries/traveling-60000km-across-chinas-belt-and-road/ (Downloaded on 22 April 2023)

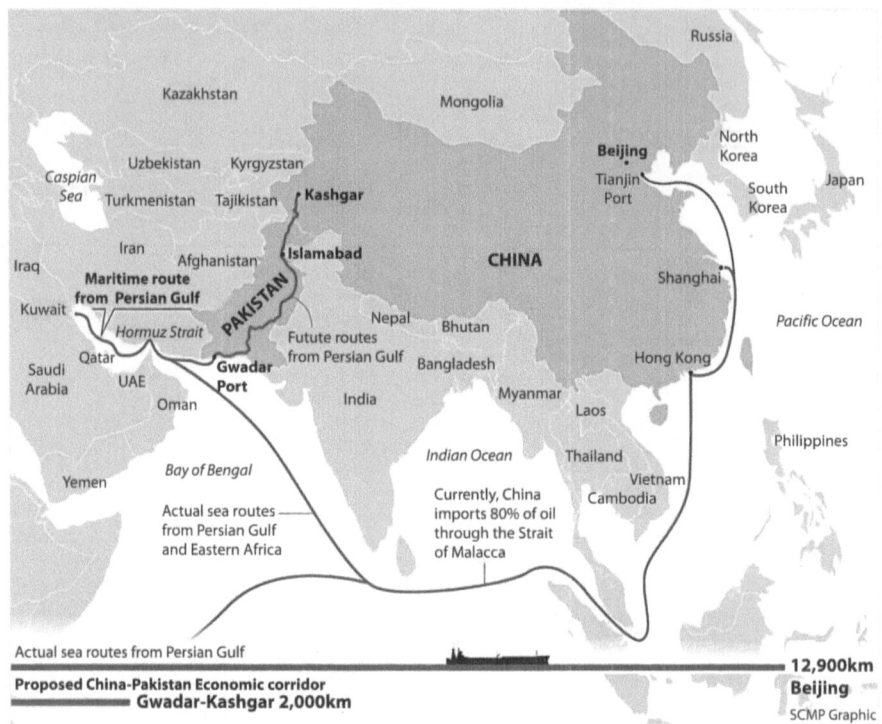

Appendix 3 Map of China Pakistan Economic Corridor (CPEC) – A short route for China's Oil supply from Persian Gulf

Source: Deloitte Study: Pakistan Economic Survey 2014–15

Available at: https://www.researchgate.net/figure/Proposed-China-Pakistan-economic-corridor-Source-Deloitte-Study-Pakistan-Economic_fig1_321726859 (Downloaded on 22 April 2023)

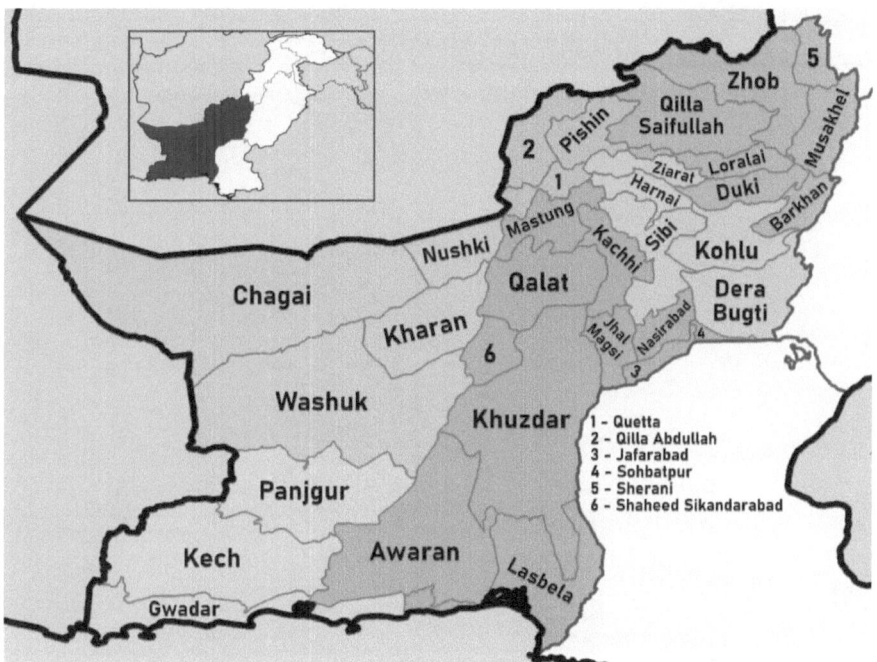

Appendix 4 Map of Balochistan Province (Pakistan) with District Names

Source: Wikimedia Commons

By Abdullah Ali Abbasi available at: https://commons.wikimedia.org/wiki/File:Districts_of_Balochistan,_Pakistan_with_district_names.png (Downloaded on 22 April 2023)

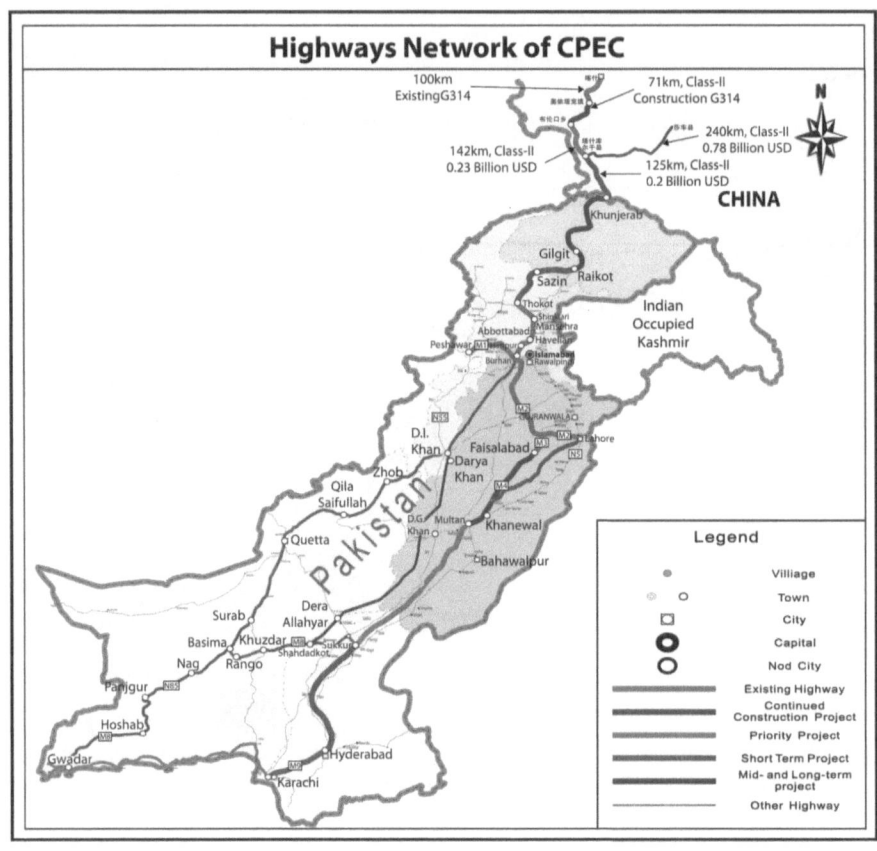

Appendix 5 Map of Highways Network (Routes) of CPEC

Source: CPEC Secretariat: Ministry of Planning, Development & Reform, Government of Pakistan. Available at: http://cpec.gov.pk/maps (Downloaded on 22 April 2023)

Appendix 6 COIN Scorecard/Table

15 Good COIN Practices	11 Bad COIN Practices
The COIN force realized at least two strategic communication factors#.	The COIN force used both collective punishment and escalating repression.
The COIN force reduced at least three tangible support factors@.	There was corrupt and arbitrary personalistic government rule.
The government realized at least one government legitimacy factorsSS.	Host-nation elites had perverse incentives to continue the conflict.
Government corruption was reduced/good governance increased since the onset of the conflict.	An external professional military engaged in fighting on behalf of the insurgents.
The COIN force realized at least one intelligence factor*.	The host nation was economically dependent on external supporters.
The COIN force was of sufficient strength to force the insurgents to fight as guerrillas.	***Fighting was initiated primarily by the insurgents.***
Unity of effort/unity of command was maintained.	***The COIN force failed to adapt to changes in adversary strategy, operations, or tactics.***
The COIN force avoided excessive collateral damage, disproportionate use of force, or other illegitimate application of force.	The COIN force engaged in more coercion or intimidation than the insurgents.
The COIN force sought to engage and establish positive relations with the population in the area of conflict.	The insurgent force was individually superior to the COIN force by being either more professional or better motivated.
Short-term investments, improvements in infrastructure or development, or property reform occurred in the area of conflict controlled or claimed by the COIN force.	The COIN force or its allies relied on looting for sustainment.
The majority of the population in the area of conflict supported or favored the COIN force.	The COIN force and government had different goals or levels of commitment.
The COIN force established and then expanded secure areas.	
Government/COIN reconstruction/development sought/achieved improvements that were substantially above the historical baseline.	
The COIN force provided or ensured the provision of basic services in areas that it controlled or claimed to control.	
The perception of security was created or maintained among the population in areas that the COIN force claimed to control.	

Source: (Paul et al., 2013a, p. 249)
Note: Practices followed by the Pakistani COIN forces in Balochistan (2013–2022) are italicized and in bold letters.

Legend: (Paul et al., 2013a; pp. 270–71)

#Strategic Communication Factors

a   COIN force and government actions were consistent with messages (delivering on promises).
b   The COIN force maintained credibility with populations in the area of conflict (includes expectation management).
c   *Messages or themes cohered with the overall COIN approach.*
d   COIN force avoided creating unattainable expectations.
e   *Themes and messages were coordinated across all involved government agencies.*

@Tangible Support Factors

a   Flow of cross-border insurgent support significantly decreased, remains dramatically reduced, or largely absent
b   Important external support to insurgents significantly reduced
c   *Important internal support to insurgents significantly reduced*
d   Insurgents' ability to replenish resources significantly diminished
e   Insurgents unable to maintain or grow force size
f   COIN force efforts resulting in increased costs for insurgent processes
g   COIN forces effectively disrupt insurgent recruiting
h   COIN forces effectively disrupt insurgent materiel acquisition
i   COIN forces effectively disrupt insurgent intelligence
j   COIN forces effectively disrupt insurgent financing

$$Government Legitimacy Factors

a   Government leaders selected in a manner considered just and fair by majority of population in area of conflict
b   Majority of citizens in the area of conflict view government as legitimate

*Intelligence Factors

a   Intelligence adequate to support kill/capture or engagements on COIN force's terms
b   *Intelligence adequate to allow COIN force to disrupt insurgent processes or operations*

Appendix 7 Details of Radio Stations in Balochistan, Pakistan

Serial	Station	Broadcasting City	Frequency	Broadcasting Languages
AM Radio Stations				
1	Radio Pakistan	Quetta	AM 750	Urdu, Balochi, Brahui, Pashto, Punjabi, Hazaragi
2	Radio Pakistan	Khuzdar	AM 560	Urdu, Balochi, Brahui, Punjabi
3	Radio Pakistan	Turbat	AM 1580	Urdu, Balochi, Punjabi
4	Radio Pakistan	Sibi	AM 1580	Urdu, Balochi, Pashto, Punjabi
FM Radio Stations				
5	Radio Chiltan	Quetta	FM 88.0	Urdu, Balochi, Brahui, Pashto, Punjabi
6	Radio Jhalawan	Khuzdar	FM 88.0	Urdu, Balochi, Brahui, Punjabi
7	Suno Pakistan	Gwadar	FM 89.4 & 96.0	Urdu, English, Punjabi, Balochi
8	Suno Pakistan	Khuzdar		
9	Suno Pakistan	Turbat		
10	Suno Pakistan	Panjgur		
11	Suno Pakistan	Sibi		
12	Radio One	Gwadar	FM 91.0	Urdu, English, Punjabi, Balochi
13	Radio Pakistan FM 101.0	Gwadar	FM 101.0	Urdu, Balochi, English, Punjabi
14	Radio Pakistan FM 101.0	Quetta	FM 101.0	Urdu, Balochi, Pashto, English, Punjabi
15	Hot FM	Quetta	FM 105.0	Urdu, Balochi, Pashto, English, Punjabi

Source: Author

Index

Note: *Italicized* pages refer figures, **bold** tables and with "n" notes in the text.